DETERMINISTIC LEARNING THEORY

FOR IDENTIFICATION, RECOGNITION, AND CONTROL

AUTOMATION AND CONTROL ENGINEERING

A Series of Reference Books and Textbooks

Series Editors

FRANK L. LEWIS, Ph.D.,
FELLOW IEEE, FELLOW IFAC

Professor
Automation and Robotics Research Institute
The University of Texas at Arlington

SHUZHI SAM GE, Ph.D.,
FELLOW IEEE

Professor
Interactive Digital Media Institute
The National University of Singapore

1. Nonlinear Control of Electric Machinery, *Darren M. Dawson, Jun Hu, and Timothy C. Burg*
2. Computational Intelligence in Control Engineering, *Robert E. King*
3. Quantitative Feedback Theory: Fundamentals and Applications, *Constantine H. Houpis and Steven J. Rasmussen*
4. Self-Learning Control of Finite Markov Chains, *A. S. Poznyak, K. Najim, and E. Gómez-Ramírez*
5. Robust Control and Filtering for Time-Delay Systems, *Magdi S. Mahmoud*
6. Classical Feedback Control: With MATLAB®, *Boris J. Lurie and Paul J. Enright*
7. Optimal Control of Singularly Perturbed Linear Systems and Applications: High-Accuracy Techniques, *Zoran Gajif and Myo-Taeg Lim*
8. Engineering System Dynamics: A Unified Graph-Centered Approach, *Forbes T. Brown*
9. Advanced Process Identification and Control, *Enso Ikonen and Kaddour Najim*
10. Modern Control Engineering, *P. N. Paraskevopoulos*
11. Sliding Mode Control in Engineering, *edited by Wilfrid Perruquetti and Jean-Pierre Barbot*
12. Actuator Saturation Control, *edited by Vikram Kapila and Karolos M. Grigoriadis*
13. Nonlinear Control Systems, *Zoran Vukiç, Ljubomir Kuljaāa, Dali Donlagiā, and Sejid Tesnjak*
14. Linear Control System Analysis & Design: Fifth Edition, *John D'Azzo, Constantine H. Houpis and Stuart Sheldon*
15. Robot Manipulator Control: Theory & Practice, Second Edition, *Frank L. Lewis, Darren M. Dawson, and Chaouki Abdallah*
16. Robust Control System Design: Advanced State Space Techniques, Second Edition, *Chia-Chi Tsui*
17. Differentially Flat Systems, *Hebertt Sira-Ramirez and Sunil Kumar Agrawal*

DETERMINISTIC LEARNING THEORY

FOR IDENTIFICATION, RECOGNITION, AND CONTROL

CONG WANG
South China University of Technology
Guangzhou, China

DAVID J. HILL
Australian National University
Act, Australia

CRC Press
Taylor & Francis Group
Boca Raton   London   New York

CRC Press is an imprint of the
Taylor & Francis Group, an **Informa** business

CRC Press
Taylor & Francis Group
6000 Broken Sound Parkway NW, Suite 300
Boca Raton, FL 33487-2742

© 2010 by Taylor and Francis Group, LLC
CRC Press is an imprint of Taylor & Francis Group, an Informa business

Library of Congress Cataloging-in-Publication Data

Wang, Cong.
    Deterministic learning theory for identification, control, and recognition / Cong Wang and David J. Hill. -- 1st ed.
        p. cm. -- (Automation and control engineering ; 29)
    Includes bibliographical references and index.
    ISBN 978-0-8493-7553-8 (alk. paper)
    1. Intelligent control systems. 2. Neural networks (Computer science) 3. Control theory. I. Hill, David J. II. Title. III. Series.

TJ217.5.W355 2009
629.8--dc22

2008038057

Visit the Taylor & Francis Web site at
http://www.taylorandfrancis.com

and the CRC Press Web site at
http://www.crcpress.com

Dedication

To our wives, Tong and Gloria.

Contents

Preface

The problem of learning in dynamic environments is important and challenging. In the 1960s, learning from control of dynamical systems was studied extensively. At that time, learning was similar in meaning to other terms such as adaptation and self-organizing. Since the 1970s, learning theory has become a research discipline in the context of machine learning, and more recently as computational or statistical learning. As a result, learning is considered as a problem of function estimation on the basis of empirical data, and learning theory has been studied mainly by using statistical principles. Although many problems in learning static nonlinear mappings have been handled successfully via statistical learning, a learning theory for dynamic systems, for example, learning of the functional system dynamics from a dynamical process, has received much less investigation.

This book emphasizes learning in uncertain dynamic environments, in which many aspects remain largely unexplored. The main subject of the monograph is knowledge acquisition, representation, and utilization in unknown dynamic processes. A deterministic framework is regarded as suitable for the intended purposes. Furthermore, this view comes naturally from deterministic algorithms in identification and adaptive control of nonlinear systems which motivate some of our work. Referred to as *deterministic learning* (DL), the learning theory presented gives promise of systematic design approaches for nonlinear system identification, dynamic pattern recognition, and intelligent control of nonlinear systems.

Deterministic Learning

The most important problem in deterministic learning is how to acquire knowledge from unknown dynamical processes. This problem is closely related to the areas of system identification and adaptive control. To achieve accurate identification of a system model, it is essential to satisfy the persistent excitation (PE) condition, which then guarantees parameter convergence in the dynamical process. Nevertheless, for identification of general nonlinear dynamical systems, the PE condition is very difficult to characterize and usually cannot be verified *a priori*.

Deterministic learning theory is mainly developed using concepts and theories of system identification, adaptive control, and dynamical systems.

Elements of the deterministic learning theory include (i) employment of the localized radial basis function network (RBFN), (ii) satisfaction of a partial PE condition along a periodic or periodic-like orbit, (iii) guaranteed exponential stability of a class of linear time-varying (LTV) adaptive systems, and (iv) locally accurate RBFN approximation of a partial system model in a local region along the periodic or periodic-like orbit. With deterministic learning, fundamental knowledge on system dynamics can be accumulated, stored, and represented by constant RBF networks in a deterministic manner. Moreover, in a scenario whereby an adaptive neural network (NN) controller achieves tracking of a periodic or periodic-like reference orbit, the deterministic learning mechanism is shown to be capable of achieving closed-loop identification of partial system dynamics during tracking control. This process implements knowledge acquisition from a closed-loop control task in uncertain dynamic environments. Different tasks will provide different knowledge (partial models of control system dynamics).

Dynamical Pattern Recognition

The problem of learning from dynamic environments is also related to the area of temporal pattern recognition. Humans generally excel in dealing with temporal patterns. Human recognition of such patterns is an integrated process in which patterns of information distributed over time can be effectively identified, represented, recognized, and classified. These recognition mechanisms, although not fully understood, are quite different from the existing conventional neural network and statistical approaches for pattern recognition. A fundamental problem in temporal pattern recognition is how to appropriately represent the time-varying patterns. This problem is difficult if a temporal pattern is to be represented in a time-independent manner. Another important problem is the characterization of similarity between two temporal patterns. As temporal patterns evolve with time, the existing similarity measures developed for static patterns do appear to be of limited usefulness.

In this book, we investigate the recognition of a class of temporal patterns generated from nonlinear dynamical systems, which are referred to as *dynamical patterns*. Based on the deterministic learning mechanism, a time-varying dynamical pattern can be effectively represented in a time-invariant and spatially distributed manner by using the locally accurate RBFN approximation of system dynamics underlying the dynamical pattern. Similarity of dynamical patterns is characterized by comparison of the system dynamics inherent within these dynamical patterns. A mechanism for rapid recognition of dynamical patterns is presented, by which a test dynamical pattern is recognized as similar to a training dynamical pattern if state estimation or synchronization is achieved according to a kind of internal and dynamical matching on system dynamics. Thus, rapid recognition of dynamical patterns is implemented due to the effective utilization of the learned knowledge in dynamic environments.

Pattern-Based Intelligent Control

Concerning the problem of knowledge acquisition and utilization in dynamic environments with feedback control, we investigate the topic of pattern-based intelligent control. This was studied tentatively in the 1960s, but not further developed to its potential. It has been a natural idea to combine pattern recognition with automatic control, which is intuitively motivated by the capabilities of human learning and control. A human can learn many highly complicated control tasks, and these tasks can then be performed repeatedly with little effort. The implementation of this idea in control technology, however, has been a big challenge. Difficulties include representation, similarity measures, and rapid recognition and classification of different control situations which are here referred to as dynamical patterns. It is obvious that conventional pattern recognition methods are not suitable to solve these problems.

In this book, we propose a framework for pattern-based intelligent control. Fundamental knowledge concerning different control situations is identified via deterministic learning. A set of training dynamical patterns is defined based on the identification. For a test control situation, if it is classified as similar to one previous training pattern, then the neural network (NN) controller corresponding to the training pattern is selected and used. This effectively exploits the learned knowledge to achieve guaranteed stability and improved control performance. The proposed pattern-based intelligent control bears similarity to proficient human learning and control. It will be useful in areas such as motion control of robotics and security assessment and control of power systems.

Organization of the Book

This book is aimed at researchers in broad areas of systems and control, such as nonlinear system identification, adaptive control, neural networks control, and temporal pattern recognition. It is also intended to be used for advanced study as the text for a graduate-level course. The results on which the book is based were reported in the literature only recently (the main ones from 2006). The book aims to expand on these and further develop the subject. Nevertheless, the results are presented at a level accessible to audiences with a standard background in concepts and theorems of dynamical systems and control.

The first chapter provides an introduction to the principal concepts of deterministic learning theory. It introduces many of the central ideas, such as satisfaction of a partial PE condition, parameter convergence, and locally accurate approximation. These are discussed at greater length in later chapters of the book. Chapter 2 is devoted to the establishment of the property of persistence of excitation (PE) for RBF networks. Chapter 3 describes the basic theory of deterministic learning processes. This includes partial parameter convergence and locally accurate approximation of nonlinear system dynamics. Chapter 4 deals with the problem of deterministic learning in closed-loop feedback control processes. Chapter 5 presents a unified framework for effective representation, similarity characterization, and rapid recognition of dynamical patterns. Chapter 6 describes pattern-based intelligent control. Chapter 7 is devoted to the practical problem of deterministic learning, where only a single output measurement is available, and to the problem of representation and rapid recognition of single-variable dynamical patterns. Chapter 8 gives conclusions and discusses some problems in deterministic learning theory for further research.

Acknowledgments

This book arose from joint work by the authors that started when both were at City University of Hong Kong. Cong Wang had completed his Ph.D. in adaptive NN control and was investigating unresolved issues toward "smart" NN control, and David Hill was exploring ways to achieve so-called global control, that is, control at several levels that can self-organize in the presence of changing goals and disturbances (of which humans are capable). They continued exploring the possibilities of NN-based learning in dynamic environments at South China University of Technology and The Australian National University, respectively. The results clearly overcame some issues left unresolved by the ARMAX and state space–based model approaches to system identification and adaptive control for nonlinear systems and the ideas behind this book emerged.

The authors would like to thank many people who have in various ways helped us complete this book. We are especially grateful to Shuzhi S. Ge, who introduced the first author to the field of adaptive NN control back in 1998, and to Guanrong Chen, who helped us lay the foundations of important parts of the book in 2003. We would like to express our deepest appreciation to Jie Huang, Frank L. Lewis, and Chenghong Wang who gave us great support in writing this book. They have been great advisors, friends, and colleagues.

We are grateful to many people for their discussions and interactions that helped us broaden our understanding of the field, including Daizhan Cheng, Nanning Zheng, Hongxin Wu, Xinghuo Yu, Xiaohua Xia, Deyi Li, Zhiyong Liu, Jie Chen, Feiyue Wang, Daren Yu, Dewen Hu, Guangren Duan, Donghua Zhou, Wenxin Qin, Jun Zhao, Xiaofeng Wu, Su Song, Changyin Sun, Yong Wang, and Zejian Yuan. The second author also thanks Peter Neilson for discussions some years ago on human body control and also benefited from even earlier collaboration on adaptive control, particularly with Changyun Wen. We are thankful to our students who collaborated with us in research and contributed to this work: Tengfei Liu, Tianrui Chen, Guopeng Zhou, Zhengui Xue, Tao Peng, and Binhe Wen.

We would also like to extend our thanks to our colleagues and friends at South China University of Technology, The Australian National University, and especially at the National Natural Science Foundation of China for their friendship, support, and technical interactions.

The first author acknowledges the support of South China University of Technology, the National Natural Science Foundation of China (under Grant No. 60743011), the program of New Century Excellent Talents in Universities

(NCET), and the 973 Program (under grant No. 2007CB311005). The second author acknowledges the support of City University of Hong Kong, the Research Grants Council of Hong Kong, The Australian National University, and the Australian Research Council during the prior work and writing of the book. Special thanks are due to the editors Nora Konopka and Theresa Delforn of Taylor & Francis for their enthusiasm and help, which made this book possible.

Finally, we thank our families, especially Tong and Gloria, for their love, encouragement, and patience, which helped greatly to ensure that we completed this book.

About the Authors

Cong Wang received both B.E. and M.E. degrees from the Beijing University of Aeronautics & Astronautics in 1989 and 1997, respectively, and a Ph.D. from the Department of Electrical & Computer Engineering, National University of Singapore in 2002. From 2001 to 2004, he did his postdoctoral research at the Department of Electronic Engineering, City University of Hong Kong. He has been with the College of Automation, South China University of Technology, Guangzhou, China, since 2004, where he is currently a professor. Dr. Wang has authored and co-authored over 40 international journal and conference papers. From May 2005 to August 2007, he worked as a program director at the Department for Information Sciences, National Natural Science Foundation of China (NSFC). He serves as an associate editor of the IEEE Control Systems Society (CSS) Conference editorial board. His research interests include deterministic learning theory, dynamical pattern recognition, pattern-based intelligent control, and cognitive and brain sciences.

David J. Hill received B.E. and B.Sc. degrees from the University of Queensland, Australia, in 1972 and 1974, respectively. He received a Ph.D. in electrical engineering from the University of Newcastle, Australia, in 1976. He is currently a professor and Australian Research Council Federation Fellow in the Research School of Information Sciences and Engineering at The Australian National University. He is also deputy director of the Australian Research Council Centre of Excellence for Mathematics and Statistics of Complex Systems. He has held academic and substantial visiting positions at the universities of Melbourne, California (Berkeley), Newcastle (Australia), Lund (Sweden), Sydney, and Hong Kong (City University). Dr. Hill holds honorary professorships at the University of Sydney, University of Queensland (Australia), South China University of Technology, City University of Hong Kong, Wuhan University, and Northeastern University (China). His research interests are in network systems science, stability analysis, nonlinear control, and applications. He is a fellow of the Institution of Engineers, Australia, the Institute of Electrical and Electronics Engineers, United States, and the Australian Academy of Science; he is also a foreign member of the Royal Swedish Academy of Engineering Sciences.

1

Introduction

The objective of this book is to present a recently developed framework for learning from uncertain dynamic environments, which allows further developments in the area of knowledge acquisition, representation, and utilization in dynamical processes. Referred to as *deterministic learning* (DL), the learning mechanism that underpins the framework provides systematic approaches for identification, recognition, and control of nonlinear dynamical systems. The book is justified by the aim to collect and expand the basic ideas and results, although there appears to be much more research needed for the topic to be fully developed.

The problem of learning in dynamical or non-stationary environments so far has received minor attention compared to the problem of learning in static or stationary environments. In this book, we investigate two types of uncertain dynamic environments: (i) feedback control of uncertain nonlinear systems, and (ii) recognition and classification of temporal/dynamical patterns. These topics are closely connected in that they are both parts of decision and control for complex situations. In this chapter, we start by revisiting different areas of feedback control concerning the problem of learning in dynamic processes. Specifically, Section 1.1 discusses the learning issues in related areas such as adaptive control, learning control, intelligent control, and adaptive neural network (NN) control. The learning issues in temporal pattern recognition are included in Section 1.2. Difficulties concerning the occurrence of learning in these dynamical processes are analyzed, respectively, in the two sections. In Section 1.3, we briefly introduce the main topics of this book, including a more detailed introduction to the above-mentioned learning issues and the basic ideas leading to the development of the deterministic learning theory.

1.1 Learning Issues in Feedback Control

1.1.1 Adaptive and Learning Control

Adaptive control has been the subject of active research for more than a half century; see some history in the well-known text by Astrom and Wittenmark [13]. According to *Webster's Dictionary*, to adapt means "to change (oneself) so that one's behavior will conform to new or changed circumstances."

The words "adaptive system" and "adaptive control" have come to refer to situations where the controller has adjustable parameters and some process for changing them as new conditions are encountered. The motivation of adaptive control was originally to design autopilots for high-performance aircraft undergoing drastic changes in their dynamics when they fly from one operating point to another. These changes could not be handled by constant-gain feedback control. However, in the 1950s there was a lack of rigorous analysis for the stability of the proposed adaptive flight control schemes. The introduction of state-space techniques and Lyapunov stability theory [103] made the 1960s an important period for the development of adaptive control theory [17]. The advances in the 1960s improved the understanding of adaptive systems and contributed to a strong renewed interest in the field in the 1970s. Since then, there have been many theoretical successes and some applications. There are too many important works to refer to here; see the surveys and books, including [5,12,13,78,92,119,125,152,159,161,199,226] for more details.

The objective of adaptive control is clearly defined and compelling: to control linear or nonlinear systems with uncertain parameters [119]. Adaptive control has as a key feature the ability to adapt to, or "learn," the unknown parameters during online adjustment of controller parameters in order to achieve a desired level of control performance. The emphasis of adaptive control theory is on the stability of adaptive systems. However, the learning ability of conventional adaptive control is actually very limited. To be specific, in the process whereby an adaptive control algorithm adjusts the controller parameters online so that closed-loop stability is maintained, one may argue that learning is achieved in the sense that the adaptive system learns enough about the system to deal with uncertain parameters. However, even for repeating exactly the same control task, the adaptive control algorithm still needs to recalculate the controller parameters because nothing was kept in memory. In this sense, the adaptive system does not have a learning capability.

Learning control also started to receive increased attention in the 1960s [15,55]. At that time, adaptation, learning, self-organizing systems, and control were competing terms having similar but somewhat undeveloped meanings. The basic idea of learning control is as follows. When information about the controlled process (plant and environment) is unknown, a controller is designed that is capable of estimating the unknown information during its operation. If the estimated information gradually approaches the true information as time proceeds, then the performance of the designed controller will eventually be as good as in the case where all the information required is known. This class of control systems may be called learning control systems because the gradual improvement of performance is due to the improvement of the estimated unknown information [56]. Here the learned information is considered as an experience of the controller, and the experience will be used to improve the quality of control whenever similar control situations recur.

From the concepts introduced, the problem of learning may be viewed as estimation or successive approximation of the unknown quantities that represent the controlled process under study. The unknown quantities to be estimated or learned by the controller may be either the parameters only, or structure of a deterministic or stochastic function. The term "learning" is unambiguously explained in terms of the appropriate utilization of past experience and the gradual improvement of performance. The difference between basic adaptive control and learning control lies in that an adaptive control system recalculates the controller parameters repeatedly without any knowledge learned and kept in memory; a learning control system requires not only the adaptive capability to cope with system uncertainties, but also other capabilities beyond that of adaptation, for example, knowledge acquisition, storage, and reuse for another similar control task [44].

Learning is clearly a very desirable characteristic of advanced control systems. For instance, in the trend toward control for more complex systems, it offers the opportunity of reduced computational burden as past experiences are exploited in similar new situations. According to *Webster's Dictionary*, to learn means "to acquire or gain knowledge or skills." A learning control system captures this idea and is one that has the following capabilities: (i) to acquire knowledge through closed-loop interactions with the plant and its environment, (ii) to store the knowledge in memory, and (iii) to reuse the learned knowledge (also called past experience) when similar control situations recur toward improved control performance. However, just to gain knowledge in a dynamical closed-loop control process, that is, learning in a nonstationary environment for nonlinear systems, is a very difficult problem [56], which has remained incompletely solved for a long period of time.

Nowadays it is interesting to notice that, although the similarities and differences between adaptive control and learning control have been clarified, the developments of the two research areas are quite different. Adaptive control has received continuing popularity since the 1970s, with a rich literature on different techniques for design, analysis, performance, and applications. Throughout the 1980s, robust adaptive control was studied intensively [92]. The objective was to understand the mechanisms of instabilities for adaptive control algorithms in the presence of unmodeled dynamics or bounded disturbances and to propose various robustness modifications. Since the late 1980s, with the publication of several breakthrough results, adaptive control of certain classes of nonlinear plants with unknown parameters has been the focus of research, and this led to a further strong interest in the field, with some successful industrial applications [119]. On the other hand, since the 1970s learning control has been merged into a more general area called intelligent control [57], which in turn is influenced by control theory and artificial intelligence. Intelligent control has since become one of the most active research areas in the field of control; however, the precise learning capabilities of intelligent control in the sense referred to above have been somewhat lightly investigated.

Another development related to learning control is learning theory. Since the 1970s, learning theory has gradually become a research discipline in the context of machine learning, and more recently has featured computational or statistical learning using stochastic principles [229]. Although statistical learning theory could provide efficient learning algorithms for a wide variety of problems in the robust analysis and synthesis of control systems (e.g., see [234]), it is difficult to apply to practical control systems, for the models are mostly dynamical and deterministic by nature. Thus, for control systems design, it is preferred to have a learning capability that can be implemented in a deterministic manner.

1.1.2   Intelligent Control and Neural Network Control

Intelligent control was originally developed to motivate discussion of several areas related to learning control, with the emphases on problem solving or high-level decision capability [57]. Compared with learning control, intelligent control is a more general term describing the intersection of the fields of automatic control systems and artificial intelligence. The motivation of intelligent control lies in the attempt by control engineers to design more and more human-like controllers with adaptation and learning capabilities. On the other hand, many research activities in artificial intelligence, including machine learning and pattern recognition, might usefully be applied to solve learning control problems. This overlap of interest between the two areas has created many points of interest for control engineers. Furthermore, it was proposed that intelligent control should analytically investigate control systems with cognitive capabilities that could successfully interact with the environment. Therefore, in the early 1980s intelligent control was considered as a fusion of research areas in systems and control, computer science, and operations research, among others [197,198].

Intelligent control systems are typically able to perform one or more of the following functions: learning from past experiences, identifying changes that threaten the system behavior, such as failures, and reacting appropriately with planning actions at different levels of detail. This identifies the areas of machine learning, neural networks (NN), fuzzy systems, failure diagnosis, and planning and expert systems, to mention but a few, as existing research areas that are related and important to intelligent control. We do not consider this further here and so do not make any attempt to relate all those areas to learning. Only one area, namely, neural networks, features strongly in the sequel.

NN control was originally inspired by the learning and control abilities of human beings, which enable them to perform with ease many complicated tasks within uncertain environments. Since the mid-1980s, control of uncertain nonlinear dynamical systems using NNs has attracted tremendous interest in the control community [82]. NNs have many features that cope with the increasing demand for controlling complex, highly uncertain, nonlinear systems in industrial applications, including highly parallel structure, learning ability, nonlinear function approximation, fault tolerance, and efficient analog VLSI implementation for real-time applications. The use of neural networks in principle makes it unnecessary to spend much effort on system modeling in cases where such modeling is difficult.

In NN control of nonlinear systems, the unknown nonlinear system dynamics are approximated by linearly or nonlinearly parameterized neural networks, such as radial basis function (RBF) networks and multilayer neural networks (MNNs) (see [64]). In the earlier NN control schemes, optimization techniques were used mainly to derive parameter adaptation laws. The NN control design was demonstrated mostly through simulation or by particular experimental examples [82]. The disadvantage of optimization-based NN controllers is that it is generally difficult to derive analytical results for stability analysis and performance evaluation of the closed-loop system [64].

To overcome these problems, adaptive NN control approaches (e.g., [26,27,65,162,163,179,181,190,191,195,216,262,266,269]) were proposed based on robust adaptive control techniques [92]. The features of adaptive NN control include: (i) the design and analysis is based on Lyapunov stability theory; (ii) stability and performance of the closed-loop control system can be readily determined; and (iii) NN weights are tuned online, using a Lyapunov synthesis method, rather than optimization techniques. It has been found that adaptive NN control is suitable for controlling highly uncertain, nonlinear, and complex systems. A great deal of progress has been made both in theory and practical applications; however, there still remain some (fundamental) issues and even criticisms to be further investigated and addressed:

1. Most of the work in the NN control literature only requires the universal function approximation capability of neural networks, which is also possessed by many other function approximators, such as polynomial, rational and spline functions, wavelets, and fuzzy logic systems. As one of the online approximation-based control methods [181], it is perhaps of concern that "neural control can be accomplished without specific references to neural networks" [163]. Therefore, a question naturally arose as to what other properties particular to neural networks should be exploited to make NN control distinct from the other control methods.

2. Because NN control, as well as other online approximation-based controls, has been developed along the lines of well-established robust adaptive control theory [92], it was soon indicated that there had been no theoretical results in the adaptive neuro–fuzzy literature that would in any way use properties particular to neural networks or fuzzy systems [214]. Furthermore, it was reasonably questioned [171] whether the works of neural/fuzzy control have contributed to the understanding of adaptive systems in general. These critical comments need to be addressed.

3. Adaptive NN control has as a main feature the ability to adapt to, or "learn" the unknown system dynamics through online adjustment of controller parameters in order to achieve a desired level for control performance. However, the learning ability of adaptive NN control is actually very limited. As described above for adaptive

control generally, it needs to recalculate (or readapt) the controller parameters even for repeating exactly the same control task [44].

As both intelligent control and NN control were initially motivated by the learning and control abilities of human beings, intelligent control including NN control should at least possess the following two properties: (1) be capable of learning "good" knowledge online through a stable closed-loop control process, and (2) be capable of exploiting the learned knowledge in the same or similar control tasks with closed-loop stability and improved control performance. Properties (1) and (2) are two basic features of advanced intelligent control systems [6,44], in which the ability to learn autonomously is one of the fundamental attributes. However, these two properties in general have not been fully implemented together in the control literature.

1.2   Learning Issues in Temporal Pattern Recognition

Humans generally excel in dealing with temporal patterns, including sounds, vision, motion, and so on. Human recognition of such patterns is an integrated process, in which patterns of information distributed over time can be effectively obtained, represented, recognized, and classified. A distinguishing feature of the human recognition process is that it takes place immediately from the beginning of sensing temporal patterns, and these patterns are directly processed on the input space for feature extraction and pattern matching [34]. So far, a great deal of progress has been made for recognition of static patterns (e.g., [19,85,95,229,254,261]); however, only limited success has been reported in the literature for rapid recognition of temporal patterns. This is probably due to the lack of investigation on learning issues in temporal pattern recognition.

1.2.1   Pattern Recognition in Feedback Control

It is interesting to notice that the term pattern recognition appeared in the control literature in the 1960s together with adaptive, learning, and self-organizing systems; see, for instance, [15,55,56,226]. In the process of learning control of an uncertain linear or nonlinear system, the learned information is considered as an experience of the controller, and the experience can be used to improve the quality of control whenever similar control situations recur. Different experiences are obtained from the information extracted from different control situations. Similar control situations may be grouped to form a class of control situations. The control situations are generally referred to as patterns. Therefore, a pattern in control was represented by a set of measurements or observations of state variables [57].

The idea of using patterns to determine control actions has been employed in limited ways in specific applications. For instance, power systems are large complex systems subjected to various disturbances, which require prompt responses called emergency controls. The amount of data available is prohibitive for online computation of feedback or manual controls. It is natural to attempt to record and classify experiences as patterns defined in terms of higher-level behaviors such as recorded in operating conditions, stability indices, and trajectory trends, for example, Lissajous figures for two-dimensional projections [25,208]. Now situations can be compared to those in the database and the control action is chosen according to the similarity with past experiences. This is similar to how human body control deals with complicated tasks by storing information about past experiences in the central nervous system [166].

The problem of classifying different control situations (i.e., patterns) is important in learning control system design. Once different classes of control situations can be classified quickly and correctly, a corresponding (optimal) controller can be selected for the various classes of control situations. However, the classification might be very difficult to be implemented. For instance, consider the measurements (called features) designated as $x_1, x_2, \ldots, x_k$. They can be represented by a k-dimensional vector X in the (feature) space Ω_X. Suppose there exist m possible pattern classes (or m classes of control situations). The function of a pattern classifier is to assign (or to make a decision about) the correct class membership to each given feature vector X. Such an operation can be interpreted as a partition of the k-dimensional space Ω_X into m mutually exclusive regions (or a mapping from the space to the decision space). One problem with such a method is that the creation of a uniform partition may yield a large number of different control situations. For the partition of a multidimensional system, there will be an exponential growth with the number of subdivisions in each dimension, so that even a modest problem can yield a huge number of control situations and require a prohibitively large amount of memory.

The above problem is due to the representation of nonstationary state variables by using a finite number of different stationary patterns, and then the utilization of conventional pattern recognition techniques to identify and classify the stationary patterns. It is obvious that conventional methods for static or stationary pattern recognition have limited capability to cope with the problem. Novel methods of pattern recognition are required for classifying nonstationary patterns in feedback control systems.

1.2.2 Representation, Similarity, and Rapid Recognition

In static pattern recognition, a pattern is usually a set of time-invariant measurements or observations represented in vector or matrix notation [19,95]. The dimensionality of the vector or matrix representation is generally kept as small as possible by using a limited yet salient feature set for purposes such as removing redundant information and improving classification performance. For example, in statistical pattern recognition, a pattern is represented by a set of d features, or a d-dimensional feature vector which yields a d-dimensional feature space. Subsequently, the task of recognition or classification is accomplished when the d-dimensional feature space is partitioned into compact and disjoint regions, and decision boundaries are constructed in the feature space that separate patterns from different classes into different regions [95,254].

For representation of temporal patterns, a popular approach is to construct short-term memory (STM) models, such as delay lines [236], decay traces [101,251], and exponential kernels [217]. These STM models are then embedded into different neural network architectures. For example, the time delay neural network (TDNN) is constructed by combining multilayer perceptrons (MLPs) with the delay line model [236]. With STM models, a temporal pattern is represented as a sequence of pattern states, and recognition of temporal patterns is quite similar to the recognition of static patterns.

Because the measurements of state variables are mostly time-varying in nature, the above framework for static patterns is not very suitable for representation of temporal patterns. A very difficult problem in temporal pattern processing is how to appropriately represent the time-varying patterns. The topic of temporal coding, and particularly using neural representations, recently has also become an important topic in neuroscience and related fields (see, e.g., [249]). Among the unresolved problems in this field, one of the most fundamental questions is how temporal patterns can be represented in a time-independent manner [34]. As indicated in [34], if the time attribute could not be appropriately dealt with, the problem of *time-independent* representation without loss of discrimination power and classification accuracy would be a very difficult task for temporal/dynamical pattern recognition.

Another important problem in temporal pattern recognition is the definition of similarity between two temporal patterns. In the literature of pattern recognition, there are many definitions for similarity of static patterns, most of which are based on distances, for example, Euclidean distance, Manhattan distance, and cosine distance [254]. To define the similarity of two dynamical patterns, the existing similarity measures developed for static patterns might become inappropriate, because when considering parameter variations, noise, and disturbances, it is of course unlikely that two temporal patterns will occur identically. For the aforementioned reasons, it appears that in the current literature there are no results on efficient representation and standard similarity definitions of temporal patterns.

Considering the general recognition process for a temporal pattern, two phases exist: the identification phase and the recognition phase. The "identification" phase involves working out the essential features of a pattern one does not recognize, whereas "recognition" means looking at a pattern and realizing that it is the same or a similar pattern to one seen earlier. The recognition phase involves the utilization of knowledge or past experiences obtained from the identification phase, and is expected to be processed at a rapid speed. Note that the human recognition process appears to take place *immediately and continuously* from the beginning of sensing temporal patterns, and temporal patterns are processed directly on the input space for feature extraction and pattern matching [34]. The rapid recognition process implies that, compared with the identification phase, there is a different mechanism in this phase in which past experiences will be utilized to achieve the rapid recognition.

1.3 Preview of the Main Topics

The subject of the monograph is knowledge acquisition, representation, and utilization in uncertain dynamical processes. In this section we briefly preview the main topics to be developed. The results are based on our recently published papers [238]–[248], with many extensions.

1.3.1 RBF Networks and the PE Condition

It has been shown in system identification and adaptive control that to achieve accurate parameter convergence and the corresponding identification of system dynamics, the persistent excitation (PE) condition is normally required to be satisfied [139,161,199]. Defined as an intrinsic property of certain signals (called "regressor" vectors) in the system, the PE condition plays a central role in adaptive system theory. Nevertheless, for identification of general nonlinear systems as well as identification in closed-loop control, the PE condition is very difficult to characterize and usually cannot be verified *a priori* [140]. The difficulties concerning the PE condition lead to the question of whether there exists a special class of nonlinear regressor vectors for which these can be overcome.

In the literature of identification and control of nonlinear systems using neural networks, various types of NN architectures have been employed. In fact, the research on neural networks has led to a proliferation of architectures, structures, and algorithms. The first question to be answered is which type of neural network is most suitable for learning from dynamic environments concerning NN identification/control. More specifically, we are interested in the problem of whether there exist certain types of neural network that can lead to the satisfaction of the PE condition. A natural idea to arise is that any property of neural networks leading to the satisfaction of the PE condition would be beneficial for NN identification/control. For this book, after comparison of alternatives, we come to the conclusion that the localized radial basis function (RBF) network is very suitable to implement prespecified learning and control capabilities due to its associated properties, including the linear-in-parameter form, the function approximation ability, the spatially localized structure, and an important property concerning the PE condition.

The investigation of the PE property of RBF networks has attracted continued efforts during the past decade [80,123,143,194]. RBF networks have been widely used in identification and adaptive control of nonlinear systems

[65,114,195], thanks to the universal function approximation ability. An RBF network can be represented in the form of a linear parametric regression, as a product of a neural weight vector and a regressor vector. The components of the regressor vector are nonlinear functions of inputs to the RBF network. In [194], it was shown that if the inputs to an RBF network coincide with the network neuron centers, then the corresponding regressor vector satisfies the PE condition. This requirement is very restrictive, because a random input in most cases will not coincide with the network neuron centers. For RBF networks with neuron centers fixed on a regular lattice, it was shown that the corresponding regressor vector is persistently exciting provided that the input variables to the RBF networks belong to certain neighborhoods of the neuron centers [80,143]. Nevertheless, theoretical analysis of the size of the neighborhoods was not given. In [123], it was proven that if the size of the neighborhoods is less than one half of the minimal distance of any two neuron centers, then the corresponding regressor vector might be persistently exciting. In addition, a class of ideal input orbits, which ensure the satisfaction of the PE condition, is characterized as periodic or ergodic trajectories visiting the limited neighborhoods of all neuron centers of the RBF network [123,143]. These results, although achieving substantial improvement compared with [194], are not yet applicable to the knowledge acquisition problem at hand, because it is possible that a random input sequence or orbit does not visit the specified neighborhood of any neuron center of the RBF network.

In Chapter 2, we investigate the PE property of RBF networks. To make the result applicable to NN identification and control, it is of interest to explore whether any periodic orbit can lead to the satisfaction of the PE condition. We prove (following [123,243]) that almost any periodic or periodic-like (recurrent) NN input trajectory, as long as it stays within the domain lattice, can lead to the desired PE property of a regressor subvector consisting of RBFs whose centers are located in a neighborhood of the input trajectory. Our proof proceeds by removing the restriction on the size of the neighborhood (as done in [123]). The PE condition obtained is referred to as a "partial" PE condition, because it is not necessary for the NN input trajectory to visit every center of the entire regular lattice upon which the RBF networks are constructed.

1.3.2   The Deterministic Learning Mechanism

The employment of neural networks for learning complex input-output mappings has stimulated many studies within the context of nonlinear systems identification (see, e.g., [162,209]). In particular, design and analysis of identification algorithms based on Lyapunov stability theory provide a general formulation for modeling, identifying, and controlling nonlinear dynamical systems using NN [46,65,97,115,143,179]. Lyapunov-based NN identification is very attractive; however, it cannot achieve accurate identification/modeling of the underlying system dynamics without the satisfaction of the (PE) condition [115,143,195].

In Chapter 3, we study a deterministic mechanism for accurate NN identification of unknown nonlinear dynamical systems undergoing periodic or periodic-like (recurrent) motions. We have from Chapter 2 that with RBF networks and periodic or periodic-like NN input orbit, a partial PE condition can be satisfied. With the partial PE property, by using a dynamical version of the localized RBF network, and a Lyapunov-based adaptation law for the neural weights of the RBF network, the identification error system consisting of the state estimation error subsystem and weight estimation error subsystem can be proved to be exponentially stable along the recurrent trajectory. For neurons whose centers are close to the trajectories, the neural weights will converge to small neighborhoods of a set of optimal values, whereas for the other neurons with centers far away from the trajectories, the neural weights are not activated and almost unchanged. Thus, sufficiently accurate identification of the unknown dynamics can be achieved within a local region along the recurrent trajectory. The knowledge gained from deterministic learning can be represented as an accurate NN approximation with constant neural weights. This knowledge can be conveniently interpreted as a partial model that models the system in the neighborhood of the task trajectory. These partial models, assembled from many previous tasks, can be very valuable to call upon in future situations.

1.3.3 Learning from Adaptive Neural Network Control

As already mentioned, to guarantee accurate parameter convergence (i.e., learning) in closed-loop adaptive control, it is required that the PE condition of some internal closed-loop signals be satisfied [161]. This is often very difficult to express in terms of the external reference signals. Although interesting results on stable neural control were obtained in [43,45,46,181,195], conditions for the satisfaction of the PE condition of internal closed-loop signals have not been fully established. The recent result of the authors [243] is used in Chapter 4 to show that the difficulty of satisfying PE in a feedback closed-loop is overcome in two steps. To demonstrate the idea, we consider tracking control of the states of a simple second-order nonlinear system to the recurrent states of a reference model. In the first step, we use adaptive NN control to achieve tracking convergence of the plant states to the recurrent reference states, so that the internal plant states become recurrent signals. In the second step, thanks to the obtained tracking convergence and the associated properties of localized RBF networks, partial PE conditions are subsequently satisfied by the regression subvector constructed out of the RBFs along the recurrent tracking orbit. With the partial PE condition satisfied, it is shown that accurate NN approximation of closed-loop system dynamics can be achieved in a neighborhood of the recurrent trajectory. Further, for more general nonlinear systems in strict-feedback form and Brunovsky form, it is shown that closed-loop identification of control system dynamics can be achieved in a local region along the recurrent tracking orbit. The locally accurate closed-loop identification is achieved via direct adaptive NN control rather than indirect adaptive NN control. Thus, a true learning ability is implemented during closed-loop control processes, and this is what we mean by "learning from direct adaptive NN control"; learning is in fact a natural capability inherent in the direct adaptive NN controllers. The learned knowledge can be utilized in another similar control task to achieve stability and improved performance.

1.3.4   Dynamical Pattern Recognition

A *dynamical pattern* is defined as a recurrent system trajectory generated from the following dynamical system:

$$\dot{x} = F(x; p), \, x(t_0) = x_0 \tag{1.1}$$

where $F(x; p) = [f_1(x; p), \ldots, f_n(x; p)]^T$ represents the system dynamics that is unknown. The class of recurrent trajectories includes periodic, quasi-periodic, almost-periodic, and even chaotic trajectories; see [206] for a rigorous definition of recurrent trajectory. The dynamical pattern defined above covers a wide class of temporal patterns studied in the literature.

For identification of dynamical patterns generated from nonlinear dynamical systems, the deterministic learning mechanism proposed in Chapter 3 can be used to achieve a locally accurate NN approximation of the underlying system dynamics $F(x; p)$ within a dynamical pattern. Through deterministic learning, fundamental information about dynamical patterns is obtained and stored as sets of constant RBF neural weights.

In Chapter 5, based on the deterministic learning mechanism, a unified, deterministic framework is presented for effective representation, similarity definition, and rapid recognition of dynamical patterns. This follows from the recent paper [244]. We show first that dynamical patterns can be effectively represented in a *time-invariant* manner using the locally accurate NN approximations of system dynamics $F(x; p)$. The representation is also *spatially distributed*, because fundamental information is stored in a large number of neurons distributed along the state trajectory of a dynamical pattern. Therefore, a dynamical pattern is represented by using complete information of both the pattern state and the underlying system dynamics. This differs markedly from statistical pattern recognition, where a pattern is represented as a point in a d-dimensional feature space using a limited number of extracted features [95,254],

Concerning the similarity definition for dynamical patterns, we look to the ideas in the qualitative analysis of nonlinear dynamical systems. The similarity between two dynamical behaviors lies in the *topological equivalence* and *structural stability* of two dynamical systems (see [206] for more discussions). This implies that the similarity of dynamical patterns is determined by the similarity of the system dynamics inherently within these dynamical patterns. Thus, we propose a similarity definition for dynamical patterns based on information from both system dynamics and pattern states: dynamical pattern

$\mathcal{A}$ is similar to dynamic pattern $\mathcal{B}$ if (i) the state of pattern $\mathcal{A}$ stays within a local region of the state of pattern $\mathcal{B}$, and (ii) the difference between the corresponding system dynamics along the state trajectory of pattern $\mathcal{A}$ is small. It is seen that the time attribute of dynamical patterns is excluded from the similarity definition.

With pattern representation and similarity definitions established, we investigate the mechanism for rapid recognition of dynamical patterns. We propose an approach for rapid recognition of dynamical patterns as follows. A set of dynamical models is constructed as dynamic representations of the training dynamical patterns, in which the constant RBF networks obtained from the identification phase are embedded. The constant RBF networks can quickly recall the learned knowledge by providing accurate approximations to the previously learned system dynamics of a training dynamical pattern. When a test pattern is presented to one of the dynamical models, a recognition error system is formed, which consists of the system generating the test pattern and the dynamical model corresponding to one of the training patterns. Without identifying the system dynamics of the test pattern, an *internal* and *dynamical* matching of system dynamics of the test and training pattern proceeds in the recognition error system. The state synchronization errors will be proven to be (approximately) proportional to the differences of system dynamics. Thus, the synchronization errors can be taken as similarity measures between the test and the training dynamical patterns. The process can be rapid because it does not require numerical computation associated with identifying the test pattern dynamics and comparison of system dynamics of the two dynamical patterns.

1.3.5 Pattern-Based Intelligent Control

The study of human movement and motor behavior, in the context of motor learning and control, has emerged as an important discipline in kinesiology, psychology and neuroscience (see, e.g., [205]). A recent interesting development in this field is to study human movement via a dynamic systems approach, which exhibits features of pattern-forming dynamical systems [108]. It is shown by experiments [108] that the control and coordination of human movements at all levels is associated with dynamic patterns.

It is thus suggested that mechanisms of pattern-based learning and control may be responsible for the proficiency of complicated human control skills. In this book, we intend to use the term "pattern-based intelligent control" to convey such human-like capabilities of acquiring information of dynamic patterns for current and later use and making decisions to achieve goals all in a dynamic process. These pattern-based intelligent control abilities, however, have been less studied by the control community. Such abilities require a rigorous definition of dynamic patterns, and solutions to problems of effective representation, rapid recognition and classification of dynamical patterns. These problems, nevertheless, are difficult to solve in the pattern recognition area.

Based on the aforementioned results on deterministic learning, in Chapter 6 we propose a framework for pattern-based control as follows. First, for different training control tasks, the closed-loop system dynamics corresponding to the training control tasks are identified via deterministic learning. A set of training dynamical patterns is defined based on the identification. The representation and similarity of closed-loop dynamical patterns are also presented. A set of pattern-based NN controllers is constructed accordingly. Second, a dynamical pattern classification system is introduced that can rapidly recognize dynamical patterns and switch quickly among the set of pattern-based NN controllers. For a test control task, if the corresponding dynamical pattern is recognized as very similar to one previous training pattern, then the NN controller corresponding to the training pattern is selected and activated. The learned knowledge in training periods, also called past experiences, and stored as a set of constant neural weights, is embedded in the NN container. By appropriately choosing the initial conditions, the selected NN controller control scheme can achieve small tracking errors and a fast convergence rate with small control gains. In this way, we achieve improved control performance using the past experiences. Furthermore, the NN controller does not need adaptation of neural weights; the neural learning controller is a low-order static controller that can be easily implemented. Thus, not only stability of the closed-loop system is guaranteed, better performance is also achieved in terms of saving time and energy.

Note that if the control task corresponds to a dynamical pattern not experienced before, the identification process (as in the first step) will be restarted. The learned knowledge will yield a new NN controller which will be added to the set of pattern-based NN controllers. Of course, the time available for such extra identification is an issue and might limit what can be achieved. The proposed pattern-based intelligent control framework will be useful to many areas, including the analysis of proficient human control with little cognitive effort.

1.3.6   Deterministic Learning Using Output Measurements

In Chapters 2 to 6, the deterministic learning mechanism is revealed under full-state measurements. Chapter 7 considers deterministic learning using only partial-state or output measurements.

First, for a class of nonlinear systems undergoing recurrent motions with only output measurements, we show that identification of the underlying system dynamics can still be achieved. Specifically, by using a high-gain observer, accurate state estimation of the recurrent system states is achieved. A partial PE condition of a regression subvector constructed out of the radial basis functions (RBFs) along the recurrent estimated state trajectory is satisfied, and accurate identification of system dynamics is achieved in a local region along the estimated state trajectory.

Second, we show that the knowledge obtained through deterministic learning can be reused in another state observation process. As high gains may yield large oscillations/variations in the presence of noise, the aim is to avoid high-gain design when possible. Because the learned knowledge stored in the constant RBF networks (RFBN) actually provides locally accurately known system dynamics, we construct an RBFN-based nonlinear observer, in which the constant RBF networks are embedded as NN approximations for system dynamics. For state estimation of the same nonlinear system as previously observed, it is shown that correct state estimation can be achieved according to the internal matching of the underlying system dynamics without using high-gain domination.

Third, the result of deterministic learning with output measurements is applicable to identification, representation, and rapid recognition of single-variable dynamical patterns. For single-variable dynamical patterns, difficulties arise not only because dynamical patterns evolve with time, but also due to the highly incomplete information available. We show that the system dynamics of a set of training single-variable dynamical patterns can be locally accurately identified through high-gain observation and deterministic learning. A single-variable dynamical pattern is represented in a time-invariant and spatially distributed manner by using information on both its estimated pattern states and its underlying system dynamics. This kind of representation is taken as a static representation. A series of RBFN-based observers is constructed within which the constant RBF networks are embedded. These RBFN-based observers are taken as dynamic representations for the corresponding training dynamical patterns.

Based on the dynamic representations, rapid recognition of a test single-variable dynamical pattern can be implemented when non-high-gain observation is achieved according to a similar internal and dynamical matching process described for rapid recognition of the full state test dynamical pattern. The non-high-gain observation errors are taken again as the measure of similarity between the test and training single-variable dynamical patterns. Nonetheless, it is noticed that most state variables of the test single-variable pattern are not available from measurement. To solve this problem, a high-gain observer is employed again to provide an accurate estimate of these state variables, so that the non-high-gain observation errors can still be computed. Thus, the role of non-high-gain observation in rapid recognition of dynamical patterns, that is, to measure the similarity on system dynamics between the test and training dynamical patterns, is more clearly revealed. The non-high-gain observation makes the differences on system dynamics explicitly unfolded.

1.3.7 Nature of Deterministic Learning

The deterministic learning theory for identification, recognition, and control is presented in Chapters 2 to 7. In Chapter 8, we further investigate the nature of deterministic learning.

Key elements of deterministic learning concerning knowledge acquisition include (i) employment of the localized radial basis function network, (ii) satisfaction of a partial PE condition along a periodic or recurrent orbit, and (iii) accurate RBFN approximation of unknown nonlinear dynamics achieved in a local region along a recurrent orbit. The nature of deterministic learning concerning knowledge acquisition is related to the exponential stability of a certain class of linear time-varying (LTV) adaptive systems. With deterministic learning, fundamental knowledge of uncertain dynamic environments can be obtained.

Apart from knowledge acquisition, another phase of deterministic learning is knowledge utilization, which is of the same importance as knowledge acquisition. The value of the acquired knowledge is manifested only through utilization of the knowledge in dynamic processes, for example, in rapid recognition of dynamical patterns, pattern-based intelligent control, and non-high-gain state observation. In these dynamical processes, the learned knowledge is utilized in a completely dynamical manner via a mechanism of internal and dynamical matching of system dynamics. This presents a new model of information processing, which we refer to as dynamical parallel distributed processing (DPDP). The nature of deterministic learning concerning knowledge utilization is related to the stability and convergence of certain classes of linear time-invariant (LTI) systems.

Although deterministic learning theory was not developed using statistical principles, the physiology of deterministic learning is similar to that of statistical learning. The physiology of statistical learning is revealed by the goal of not solving the problem of estimating the values of a function at given points by estimating the entire function [229]. Similarly in deterministic learning, accurate identification of a system model is achieved only in a local region along the experienced trajectory. This physiology coincides with the essence of human intelligence. Moreover, because "intelligence" means "the capacity to acquire and apply knowledge" (according to *Webster's Dictionary*), it is seen that the deterministic learning theory presents a unified framework for knowledge acquisition and knowledge utilization in dynamical processes, and thus provides a promising new direction to implement more advanced intelligence in uncertain dynamic environments.

2

RBF Network Approximation and Persistence of Excitation

The learning issues discussed in Chapter 1 are challenging problems. In the areas of identification and adaptive control of nonlinear systems, the persistant excitation (PE) condition is normally difficult to be verified *a priori*. Although various types of neural networks have been employed to exploit the universal function approximation ability, the learning capability (i.e., accurate convergence of neural weights) in the process of closed-loop identification and control has not typically been a close consideration. Accurate parameter convergence of neural weights relies on the satisfaction of the PE condition. A question naturally arises as to whether there exist certain types of neural network that can more easily enable satisfaction of the PE condition and turn out to be more suitable for learning from dynamical environments.

In this chapter, we study the property of persistence of excitation for localized radial basis function (RBF) networks. RBF networks have received much attention during the past two decades and been widely used in identification and adaptive control of nonlinear systems due to the universal function approximation ability. It is noticed that the investigation of the PE property of RBF networks also has attracted continued efforts [80,123,143,194]. These results have achieved considerable progress; however, they are not yet applicable in practice. Therefore, it is necessary to investigate further whether RBF approximators have useful PE properties that are applicable to practical NN identification and control.

Radial basis functions have their origins in the study of multivariate approximation theory, particularly in the area of strict multivariate interpolation. In Section 2.1, we briefly introduce the concepts and theorems on RBF approximation and RBF networks. The concepts of persistence of excitation and theorems of exponential stability are included in Section 2.2. In Section 2.3, based on previous results on the PE property of RBF networks [123], we show that for almost every periodic orbit, there always exists an RBF subvector consisting of RBFs centered in a certain neighborhood of the orbit such that a partial PE condition is satisfied. This result is then extended to periodic-like trajectories generated from general nonlinear systems, which include quasi-periodic, almost-periodic, and chaotic trajectories. Therefore, almost any periodic or periodic-like orbit will lead to the satisfaction of a partial PE condition of the corresponding RBF subvector. This property makes the localized RBF network most suitable for learning in dynamic environments among the various neural network (NN) architectures.

2.1 RBF Approximation and RBF Networks

2.1.1 RBF Approximation

Approximation theory has undergone major advances during the past two decades. Fundamental approximation theory includes interpolation, least squares, and Chebyshev approximation by polynomials, splines, and orthogonal-polynomials, which are still important and interesting topics. Nonetheless, some significant developments have emerged, which include new approximating tools, nonlinear approximation, and multivariate approximation [31]. RBF approximation is one of the most often applied approaches for multivariate approximation in modern approximation theory and has been considered in many applications [23].

The problem of multivariate function approximation is: given data in n dimensions consisting of data sites $\Xi \in R^n$ and function values $f_\Xi = f(\Xi) \in R$, seek an approximant $g: R^n \rightarrow R$ to the function $f: R^n \rightarrow R$ [23]. The function f is usually unknown, but the existence and some smoothness of f normally have to be required for the purpose of analysis.

In the literature, there are various ways to find approximant $g \in G$ (where G is a linear space of approximants) to approximate f. By using radial basis functions, the approximation can take place by means of interpolation. An interpolation problem is: given a set of data pairs $\{(\xi_i, y_i)|\xi_i \in R^n, y_i \in R, i = 1, \ldots, m\}$ where ξ_i are distinct points, find a suitable function $g(x): R^n \rightarrow R$ such that for each i, $g(\xi_i) = y_i$.

For RBF approximation, the approximant g is usually a finite linear combination of translates of a radially symmetric basis function $\phi(\| \cdot \|)$:

$$g(x) = \sum_{i=1}^{m} w_i \phi(\|x - \xi_i\|), \qquad x \in R^n \qquad (2.1)$$

where $\| \cdot \|$ is the Euclidean norm and w_i are real coefficients. Radial symmetry means that the value of the function only depends on the Euclidean distance $\| \cdot \|$, and any rotation will not change the function value.

Substituting the condition for interpolation yields $y = Aw$ where y, w are vectors of y_i, w_i, respectively, and the interpolation matrix is given by

$$A = \begin{bmatrix} \phi(\|\xi_1 - \xi_1\|) & \cdots & \phi(\|\xi_1 - \xi_m\|) \\ \vdots & \ddots & \vdots \\ \phi(\|\xi_m - \xi_1\|) & \cdots & \phi(\|\xi_m - \xi_m\|) \end{bmatrix} \qquad (2.2)$$

One of the main results in RBF approximation is that the interpolation matrix A is nonsingular (sometimes even positive definite) for certain types of radial basis functions provided ξ_i are distinct points. The principal concepts that are useful to show nonsingularity of the interpolation matrix are positive definite functions and completely monotone functions [23].

DEFINITION 2.1
A function $f: R^n \to R$ is said to be *semi-positive definite* if for any set of points $\xi_1, \xi_2, \ldots, \xi_m$ in R^n the $m \times m$ matrix $A_{ij} = f(\xi_i - \xi_j)$ is nonnegative definite, that is, $c^T Ac = \sum_{i=1}^{m} \sum_{j=1}^{m} c_i c_j A_{ij} \geq 0$ for all $c = [c_1, \ldots, c_m]^T \in R^m$. If $c^T Ac > 0$ whenever the points ξ_i are distinct and $c \neq 0$, then f is *positive definite*.

DEFINITION 2.2
A function is said to be *completely monotone* on $[0, \infty)$ if, (i) $f \in C[0, \infty)$, (ii) $f \in C^\infty(0, \infty)$, and (iii) $(-1)^k f^k(t) \geq 0$ for all $t > 0$ and for all $k = 0, 1, 2, \ldots$.

The Bernstein–Widder theorem gives a characterization of the class of completely monotone functions. This theorem states that a function is completely monotone if and only if it is the Laplace transform of a nonnegative bounded Borel measure [256].

THEOREM 2.1 (**Bernstein–Widder Representation**)
A function $f : [0, \infty) \to [0, \infty)$ is completely monotone, iff it is given in the following form,

$$f(t) = \int_0^\infty e^{-t\rho} d\beta(\rho) \tag{2.3}$$

where $d\beta(\rho)$ is a finite, nonnegative Borel measure on $[0, \infty)$.

With the results on positive definite functions and completely monotone functions, the Schoenberg theorem was established in [201].

THEOREM 2.2 (**Schoenberg Theorem**)
If ϕ is completely monotone but not constant on $[0, \infty)$, then the function $\xi \mapsto \phi(\|\xi\|^2)$ is a radial, positive function on any inner product space. Thus, for any m distinct points $\xi_1, \xi_2, \ldots, \xi_m$ in such a space, the matrix $A_{ij} = \phi(\|\xi_i - \xi_j\|^2)$ is positive definite (and therefore nonsingular).

Commonly used RBFs satisfying the Schoenberg theorem include the Gaussian function, and Hardy's inverse multiquadric function [83,84]. The Gaussian function is

$$\phi(\|x - \xi_i\|) = \exp\left[\frac{-(x - \xi_i)^T(x - \xi_i)}{\eta^2}\right] \tag{2.4}$$

where $\xi_1, \ldots, \xi_q$ are distinct centers and η is the width of the receptive field.

The inverse Hardy's multiquadric function [151] is

$$\phi(x) = \frac{1}{\sqrt{\sigma_i^2 + (x - \xi_i)^T(x - \xi_i)}} \tag{2.5}$$

Both the Gaussian function and inverse multiquadric function are localized radial basis functions in the sense that $\phi(\|x - \xi_i\|) \to 0$ as $\|x\| \to \infty$.

There are other functions that are not included in Schoenberg's Theorem, for example, Hardy's multiquadric function [83,84]:

$$\phi(x) = \sqrt{\sigma_i^2 + (x - \xi_i)^T(x - \xi_i)} \tag{2.6}$$

which are also useful for interpolation in geophysics. For this case, the Micchelli theorem [151] was established as follows.

THEOREM 2.3 **(Micchelli Theorem)**
Let $\phi: [0, \infty) \to [0, \infty)$. If the derivative of ϕ is completely monotone but not constant on $[0, \infty)$, then, for any n distinct points $\xi_1, \xi_2, \ldots, \xi_m$ in a real inner-product space, the matrix $A_{ij} = \phi(\|\xi_i - \xi_j\|^2)$ is nonsingular.

The above theorems provide a rich source of RBFs that are suitable for interpolation of data in Euclidean spaces [23,31]. From our point of view, the results on the nonsingularity property of the RBF interpolation matrix A are interesting, because they provide insights into the establishment of the conditional nonsingularity of another RBF interpolation matrix (given in Equation [2.24]). This conditional nonsingularity, in turn, is essential in proving the partial PE property of RBF networks in Section 2.3.

2.1.2   RBF Networks

From the 1980s, neural networks were constructed and empirically demonstrated (using simulation studies) to approximate quite well nearly all functions encountered in practical applications. The results by Funahashi [58], Cybenko [35], and Hornik, Stinchcombe, and White [91] proved that neural networks are capable of universal approximation in a very precise and satisfactory sense. These results lead the study of neural networks from its empirical origins to a mathematical discipline. The NN approximation problem can be stated following the definition of function approximation [189].

DEFINITION 2.3 **(Function Approximation)**
If $f(x): R^n \to R$ is a continuous function defined on a compact set Ω, and $f_{nn}(W, x): R^n \times R^n \to R$ is an approximating function that depends continuously on W and x, then the approximation problem is to determine the optimal parameters W^*, for some metric (or distance function) d, such that

$$d(f_{nn}(W^*, x), f(x)) \le \epsilon \tag{2.7}$$

for an acceptably small ϵ.

To approximate the unknown function $f(x)$ by using neural networks, the approximating function $f_{nn}(W, x)$ is first chosen. The neural network weights W are then adjusted by a training set. Thus, there are two distinct problems in NN approximation [85], namely, the *representation problem*, which deals with the selection of the approximating function $f_{nn}(W, x)$, and the *learning problem*, which is to find the training method to ensure that the optimal neural network weights W^* are obtained.

RBF network models were developed by Broomhead and Lowe [22] and Poggio and Girosi [178] in the late 1980s. They were motivated by the locally tuned response observed in biological neurons, for example, in the visual or auditory systems, and developed by introducing a number of modifications to overcome the restrictions in exact RBF interpolation. Now the RBF network model has become one of the most often used NN models in the neural network literature.

The RBF networks can be considered as two-layer networks in which the hidden layer performs a fixed nonlinear transformation with no adjustable parameters; that is, the input space is mapped into a new space. The output layer then combines the outputs in the latter space linearly. Therefore, they belong to a class of linearly parameterized networks, and can be described in the following form:

$$f_{nn}(Z) = \sum_{i=1}^{N} w_i s_i(Z) = W^T S(Z) \tag{2.8}$$

where $Z \in \Omega_Z \subset R^q$ is the input vector, $W = [w_1, w_2, \ldots, w_N]^T \in R^N$ is the weight vector, $N > 1$ is the NN node number, and $S(Z) = [s_1(\|Z - \xi_1\|), \ldots, s_N(\|Z - \xi_N\|)]^T$, is the regressor vector, with $s_i(\cdot)$ being a radial basis function, and ξ_i $(i = 1, \ldots, N)$ being distinct points in state space (termed centers).

It has been proven in [174] that an RBF network (2.8), with sufficiently large node number N and appropriately placed node centers and variances, can approximate any continuous function $f(Z) : \Omega_Z \rightarrow R$ over a compact set $\Omega_Z \subset R^q$ to arbitrary accuracy according to

$$f(Z) = W^{*T} S(Z) + \epsilon(Z), \quad \forall Z \in \Omega_Z \tag{2.9}$$

where W^* are the ideal constant weights, $\epsilon(Z)$ is the approximation error ($\epsilon(Z)$ is denoted sometimes as ϵ to simplify the notation). It is normally assumed that the ideal weight vector W^* exists such that $|\epsilon(Z)| < \epsilon^*$ (with $\epsilon^* > 0$) for all $Z \in \Omega_Z$. The ideal weight vector W^* is an "artificial" quantity required for analytical purposes, and is defined as the value of W that minimizes $|\epsilon|$ for all $Z \in \Omega_Z \subset R^q$; that is,

$$W^* \stackrel{\triangle}{=} arg \min_{W \in R^N} \left\{ \sup_{Z \in \Omega_Z} \left| f(Z) - W^T S(Z) \right| \right\} \tag{2.10}$$

An important class of RBF networks for our purpose is localized RBF networks, where each basis function can only locally affect the network output.

The localized representation means that for any point Z_p in the compact set Ω_Z, we have:

$$f(Z_p) = W_p^{*T} S_p(Z_p) + \epsilon_p \tag{2.11}$$

where ϵ_p is the approximation error, and can be expressed in an order term as $O(\epsilon)$, where $O(\cdot)$ denotes the large order function $S_p(Z_p) = [s(\|Z_p - \xi_{j_1}\|),$ $\dots, s(\|Z_p - \xi_{j_p}\|)]^T \in R^{N_p}$ is a subvector of $S(Z)$ in (2.8), with $|s(\|Z_p - \xi_{j_i}\|)| > \iota$ holding for those neurons centered in an ε-neighborhood of the point Z_p; that is, $\|Z_p - \xi_{j_i}\| < \varepsilon$ ($j_i = j_1, \dots, j_p$), where $\varepsilon > 0$, ι is a small positive constant, and $W_p^* = [w_{j_1}^*, \dots, w_{j_p}^*]^T$ is a subvector of the neural weights. Equation (2.11) means that at a specific point Z_p, the smooth function $f(\cdot)$ can be approximated by using neurons located at the ε-neighborhood of this point.

Similarly, for any bounded trajectory $Z(t)$ ($\forall t \geq 0$) within the compact set Ω_Z, $f(Z)$ can be approximated using neurons located in a local region (i.e., an ε-neighborhood) along this trajectory:

$$f(Z) = W_\zeta^{*T} S_\zeta(Z) + \epsilon_\zeta \tag{2.12}$$

where $\epsilon_\zeta = O(\epsilon)$ is the approximation error, $O(\epsilon)$, $W_\zeta^* = [w_{j_1}^*, \dots, w_{j_{N_\zeta}}^*]^T \in R^{N_\zeta}$, with $N_\zeta < N$, $S_\zeta(Z) = [s(\|Z - \xi_{j_1}\|), \dots, s(\|Z - \xi_{j_{N_\zeta}}\|)]^T \in R^{N_\zeta}$, with the integers j_i defined by $|s(\|Z_p - \xi_{j_i}\|)| > \iota$ hold for some $Z_p \in Z(t)$, where ι is a small positive constant. This is true if $\|Z(t) - \xi_{j_i}\| < \varepsilon$ for some $t > 0$ and $\varepsilon > 0$.

We show that localized RBF networks have the spatially localized learning capabilities of representation, storage, and adaptation. For the localized regressor functions $S(\cdot)$ used in the adaptive law (to be designed), only neurons with centers close to the input trajectory $Z(t)$ will be activated. The adaptation in one part of the input space does not significantly affect learning and storage in a different area. The two issues are discussed further in the following sections.

Among the localized RBF networks, we use the Gaussian RBF network in the following theoretical analysis and simulations. For the Gaussian RBF network, an interesting result from [123, Corollary 4.2] provides an upper bound on the Euclidean norm of vector $S(Z)$. It states the following.

LEMMA 2.1

[123] *Consider the Gaussian RBF network (Equations [2.8] and [2.4]). Let $h = \frac{1}{2} \min_{i \neq j} \|\xi_i - \xi_j\|$, and let q and η be as in Equations (2.8) and (2.4). Then we may take an upper bound of $\|S(Z)\|$ as*

$$\|S(Z)\|^2 \leq \sum_{k=0}^{\infty} 3q(k+2)^{q-1} e^{-2hk^2/\eta^2} := s^* \tag{2.13}$$

REMARK 2.1
It can be easily proven that the sum $\sum_{k=0}^{\infty} 3q(k+2)^{q-1}e^{-2hk^2/\eta^2}$ has a limited value, because the infinite series $\{3q(k+2)^{q-1}e^{-2hk^2/\eta^2}\}$ $(k = 0, \ldots, \infty)$ is convergent by the Ratio Test Theorem [39].

Apart from the above properties, the most important reason we use the localized RBF network is due to an essential property concerning the satisfaction of the PE condition.

2.2 Persistence of Excitation and Exponential Stability

Persistence of excitation is of great importance in adaptive systems. The concept was first introduced in the context of system identification by Astrom and Bohlin [9] to express the idea that the input signal to the plant should be sufficiently rich such that all the modes of the plant are excited [263], and convergence of the model parameters is achieved. Later on in the research on adaptive control in the 1970s, it was realized that the concept of PE also played an important role in the convergence of the controller parameters to their desired values [1,153]. However, there came the question of establishing PE of some internal signals rather than external signals of adaptive control systems.

The properties related to PE have been studied in depth (see, for instance, [20,160,161,199] and the references therein). The definitions of the PE condition are as follows [161,199].

DEFINITION 2.4
A piecewise-continuous, uniformly bounded, vector-valued function $S : [0, \infty) \rightarrow R^m$ is said to satisfy the persistent excitation condition, if there exist positive constants α_1, α_2, and T_0 such that

$$\alpha_1 I \leq \int_{t_0}^{t_0+T_0} S(\tau)S(\tau)^T d\tau \leq \alpha_2 I, \qquad \forall t_0 \geq 0 \qquad (2.14)$$

where $I \in R^{m \times m}$ is the identity matrix.

According to this definition, the PE condition requires that the integral of the semidefinite matrix $S(t)S(t)^T$ be positive definite over an interval of length T_0. It is noted that if S is persistently exciting for the time interval $[t_0, t_0 + T_0]$, it is PE for any interval of length $T_1 \geq T_0$ [161]. The PE condition can also be defined and expressed in a scalar form as follows [199].

DEFINITION 2.5
A piecewise-continuous, uniformly bounded, vector-valued function $S: [0, \infty) \rightarrow R^m$ is said to satisfy the persistent excitation condition, if there

exist positive constants α_1, α_2, and T_0 such that

$$\alpha_1 \leq \int_{t_0}^{t_0+T_0} |S(\tau)^T c|^2 d\tau \leq \alpha_2, \qquad \forall t_0 \geq 0, \qquad \|c\| = 1 \qquad (2.15)$$

holds for all unit vectors $c \in R^m$.

The condition above implies that the vector $S(t)$ has a finite projection along any unit vector c over a finite interval of time.

The following definition of the PE condition is presented in [123], which is suitable for RBF network identification in continuous and discrete time cases.

DEFINITION 2.6

Let μ be a positive, Σ-finite Borel measure on $[0, \infty)$. A continuous, uniformly bounded, vector-valued function $S: [0, \infty) \to R^m$ is persistently exciting, if there exist positive constants α_1, α_2, and T_0 such that

$$\alpha_1 \|c\|^2 \leq \int_{t_0}^{t_0+T_0} |S(\tau)^T c|^2 d\mu(\tau) \leq \alpha_2 \|c\|^2, \qquad \forall t_0 \geq 0 \qquad (2.16)$$

holds for every constant vector $c \in R^m$.

The definitions above reveal that PE can be defined as an intrinsic property of a class of signals. This property is closely related to the exponential stability of a class of linear time-varying systems. We first summarize some well-known stability definitions [111].

DEFINITION 2.7

Consider the system

$$\dot{x} = f(x, t), \qquad x(t_0) = x_0 \qquad (2.17)$$

where $f : [0, \infty) \times D \to R^n$ is piecewise continuous in t and locally Lipschitz in x on $[0, \infty) \times D$ where $D \in R^n$. The solution of system (2.17) starting from initial condition (t_0, x_0) is denoted as $x(t; t_0, x_0)$.

The equilibrium point $x = 0$ of system (2.17) is stable if for every $\epsilon > 0$, there exists a $\delta(\epsilon, t_0) > 0$ such that $\|x_0\| < \delta$ implies that $\|x(t; t_0, x_0)\| < \epsilon$ $\forall t \geq t_0$. It is *uniformly stable* (u.s.) if δ is independent of t_0.

The equilibrium point $x = 0$ is uniformly asymptotically stable (UAS) if it is uniformly stable and for some $\epsilon_1 > 0$ and every $\epsilon_2 > 0$, there exists $T(\epsilon_1, \epsilon_2) > 0$ such that if $\|x_0\| < \epsilon_1$, then $\|x(t; t_0, x_0)\| < \epsilon_2$ for all $t \geq t_0 + T$.

The equilibrium point $x = 0$ is exponentially stable if there exist constants a, b and $c > 0$ such that $\|x(t; t_0, x_0)\| \leq a e^{-b(t-t_0)} \|x_0\|$, for all $t \geq t_0$ and $\|x(t_0)\| < c$.

The equilibrium point $x = 0$ is uniformly exponentially stable (UES) if there exist constants $a, b > 0$ and $r > 0$ such that for all $(t_0, x_0) \in [0, \infty) \times B_r$ where $B_r = \{x \in R^n | \|x\| \leq r\}$, $\|x(t; t_0, x_0)\| \leq a e^{-b(t-t_0)} \|x_0\|$ for all $t \geq t_0$.

It is uniformly globally exponentially stable (UGES) if there exist constants $a, b > 0$ such that for all $(t_0, x_0) \in [0, \infty) \times R^n$, $\|x(t; t_0, x_0)\| \le a e^{-b(t-t_0)} \|x_0\|$ for all $t \ge t_0$.

The solution of system (2.17) is uniformly bounded if there exists a constant $c > 0$, and for every $a \in (0, c)$, there is a constant $b > 0$, independent of t_0, such that $\|x(t_0)\| \le a \Rightarrow \|x(t)\| \le b, \forall t \ge t_0$.

As an indication of the usefulness of PE to system identification and adaptive control of linear or nonlinear systems, we state the following result on exponential stability of a class of linear time-varying (LTV) systems. This problem was studied simultaneously in [1,153,263] and nicely summarized in [5,92,161,199]. The LTV system arises as the equations describing the whole adaptive system where $S(t)$ refers to the so-called regressor vector.

THEOREM 2.4
Consider the LTV system

$$\begin{bmatrix} \dot{x}_1 \\ \dot{x}_2 \end{bmatrix} = \begin{bmatrix} A & B S(t)^T \\ -S(t)C^T & 0 \end{bmatrix} \begin{bmatrix} x_1 \\ x_2 \end{bmatrix} \tag{2.18}$$

where $x_1 \in R^n$, $x_2 \in R^m$, $x = [x_1^T, x_2^T]^T \in R^{n+m}$ is the system state. If (i) the triple (A, B, C) is strictly positive real, that is, if there exist symmetric positive definite matrices P, Q, such that $PA + A^T P = -Q, PB = C$ hold,[1] and (ii) $S(t)$ is continuous and bounded and $\dot{S}(t)$ is bounded, and $S(t)$ is persistently exciting, then $x = 0$ of system (2.18) is uniformly globally exponentially stable.

For more general LTV systems in the following form:

$$\begin{bmatrix} \dot{x}_1 \\ \dot{x}_2 \end{bmatrix} = \begin{bmatrix} A(t) & B^T(t) \\ -C(t) & 0 \end{bmatrix} \begin{bmatrix} x_1 \\ x_2 \end{bmatrix} \tag{2.19}$$

where $x_1 \in R^n$, $x_2 \in R^m$, $A(t) \in R^{n \times n}$, $B(t) \in R^{m \times n}$, and $C(t) \in R^{m \times n}$, sufficient and necessary conditions for exponential stability of system (2.19) were studied in [173,268].

ASSUMPTION 2.1
[173] There exists a $\phi_M > 0$ such that, for all $t \ge 0$, the following bound is satisfied

$$\max \left\{ \|B(t)\|, \left\| \frac{d B(t)}{dt} \right\| \right\} \le \phi_M \tag{2.20}$$

---

[1] This is referred to as the Kalman–Yakubovich–Popov (KYP) lemma; see [111] and the references therein.

ASSUMPTION 2.2

[173] The system $\dot{x} = A(t)x$ is uniformly globally exponentially stable.

ASSUMPTION 2.3

[173] There exist symmetric matrices $P(t)$ and $Q(t)$ such that $A^T(t)P(t) + P(t)A(t)$ + $\dot{P}(t) = -Q(t)$ and $P(t)B(t) = C(t)$. Furthermore, $\exists\ p_m, q_m, p_M,$ and $q_M > 0$ such that $p_m \leq P(t) \leq p_M$ and $q_m \leq Q(t) \leq q_M$.

THEOREM 2.5

[173] The system (2.19) under Assumption (2.1), Assumption (2.2) and Assumption (2.3) is uniformly globally exponentially stable if and only if $B(t)$ satisfies the PE condition.

REMARK 2.2

The above two theorems establish the relationship between the PE condition and the exponential stability of two classes of LTV systems. The exponential stability of LTV systems can lead to accurate parameter convergence and system identification, which are elements of the deterministic learning mechanism introduced in the following chapters. Thus, it will be revealed that the nature of this deterministic learning mechanism is related to the exponential stability of LTV systems, which is caused by the satisfaction of the PE condition.

The following result states the robustness property of nominal systems with exponential stability (see [111] and the references therein). It shows that if the nominal system is perturbed by an arbitrarily small (or uniformly bounded) disturbance, the solution of the perturbed system will be ultimately bounded by a small bound.

THEOREM 2.6

Consider the system

$$\dot{x} = f(x, t) + g(x, t) \tag{2.21}$$

where $f : D \times [0, \infty) \rightarrow R^n$ and $g : D \times [0, \infty) \rightarrow R^n$ are piecewise continuous in t and locally Lipschitz in x on $[0, \infty) \times D$ where $D \in R^n$.

Let $x = 0$ be an exponentially stable equilibrium point of the normal system (2.17). Suppose the term $g(x, t)$ is uniformly bounded by a positive constant δ; that is, $\|g(x, t)\| < \delta$ for all $t \geq 0$ and all $x \in D$. Then, the solution of system (2.21) is uniformly bounded, that is, $\|x(t)\| < b$ for all $t \geq T$, where T is finite, and b is proportional to δ.

REMARK 2.3

This result enables statements of stability for systems such as Equations (2.18) and (2.19) to hold robustly, that is, in the presence of model imperfections [92]. Again, this facility is important in the sequel.

2.3   PE Property for RBF Networks

The PE property of RBF networks has been studied over the past decade [80,123,143,194]. One of the early attempts shows that if the inputs to an RBF network coincide with the network neuron centers, then the corresponding regressor vector satisfies the PE condition [194]. This requirement is very restrictive, because a random input in most cases will not coincide with the network neuron centers. For RBF networks with neuron centers fixed on a regular lattice, it was shown that the corresponding regressor vector is persistently exciting provided that the input variables to the RBF networks belong to certain neighborhoods of the neuron centers [80,143]. Nevertheless, theoretical analysis of the size of the neighborhoods was not provided.

An interesting result on the PE property of RBF approximants was given by Kurdila, Narcowich, and Ward [123], which shows that the regressor vector constructed out of RBF approximants is persistently exciting provided a kind of "ergodic" condition is satisfied. The size of the neighborhoods is restricted to be less than one half of the minimal distance between any two neuron centers, and a class of ideal input trajectories, which ensure the satisfaction of the PE condition, are characterized as periodic or ergodic trajectories visiting the limited neighborhoods of all neuron centers of the RBF network.

These results have achieved significant progress compared with [194], nevertheless, they are not yet applicable in practice, because it is possible that a random input sequence or orbit does not visit the specified neighborhood of all neuron centers of the RBF network. Therefore, it is necessary to investigate whether any periodic orbit can lead to the satisfaction of the PE condition.

In this section, we establish a property of persistence of excitation that is applicable for NN identification and control design. Some results presented in this section are based on the authors' papers [242,243].

When RBF networks are employed in NN identification and control, the regressor vector $S(Z(t))$ has the form

$$S(Z(t)) = [s(\|Z(t) - \xi_1\|), \ldots, s(\|Z(t) - \xi_N\|)]^T \qquad (2.22)$$

where $s(\cdot)$ is a radial basis function, ξ_i ($i = 1, \ldots, N$) are distinct points in the state space and are termed as centers, and $Z(t)$ is the state trajectory which is taken as the NN input. The function $Z(t)$ is a continuous map from $[0, \infty)$ to R^n, and it is normally assumed to be bounded in a subset of R^n.

In the following, we revisit the results on the PE property in [123]. Two interesting lemmas are given first.

LEMMA 2.2

[123] *Let $c \in R^n$ and let $Z \in R^n$ be fixed. For localized RBFs, $s(\cdot)$ satisfying (2.3),*

$$\left| \sum_{j=1}^{N} s(\|Z(t) - \xi_j\|)c_j \right|^2 \leq \left(\sum_{j=1}^{N} s(\|Z(t) - \xi_j\|)^2 \right) \|c\|^2 \leq s(o)^2 N \|c\|^2 \quad (2.23)$$

LEMMA 2.3

[123] *Let $Z_i \in R^n$ for $i = 1, \ldots, N$. If*

$$A = A(Z_1, \ldots, Z_N)$$

$$= \begin{bmatrix} s(\|Z_1 - \xi_1\|) & \cdots & s(\|Z_1 - \xi_N\|) \\ \vdots & \ddots & \vdots \\ s(\|Z_N - \xi_1\|) & \cdots & s(\|Z_N - \xi_N\|) \end{bmatrix} \quad (2.24)$$

where $s(\cdot)$ is an RBF of the form (2.8), then there exist a number $\varepsilon > 0$ and a number $\theta = \theta(\varepsilon, \xi_1, \ldots, \xi_N) > 0$ such that

$$\|Ac\| \geq \theta \|c\| \quad (2.25)$$

holds for all $c \in R^N$ and for all sets of Z_i satisfying $\|Z_i - \xi_i\| \leq \varepsilon$ for $i = 1, \ldots, N$.

The proof of Lemma 2.3 is included below for completeness of representation.

PROOF   Let $\lambda(Z_1, \ldots, Z_N)$ be the smallest eigenvalue of $A(Z_1, \ldots, Z_N)^T$ $A(Z_1, \ldots, Z_N)$, whose components are real continuous functions of $Z_1, \ldots, Z_N$. It is clear that θ^2 is a lower estimate of $\lambda(Z_1, \ldots, Z_N)$. It is also seen that $\lambda(Z_1, \ldots, Z_N)$ is a continuous function of $Z_1, \ldots, Z_N$. As $A(\xi_1, \ldots, \xi_N)$ is nonsingular, $\lambda(\xi_1, \ldots, \xi_N) > 0$. Therefore, one may choose $\varepsilon > 0$ so that $\lambda(Z_1, \ldots, Z_N) > \frac{1}{2}\lambda(\xi_1, \ldots, \xi_N) > 0$ holds for Z_i satisfying $\|Z_i - \xi_i\| \leq \varepsilon$, $i = 1, \ldots, N$. Choosing $\theta = \sqrt{\frac{1}{2}\lambda(\xi_1, \ldots, \xi_N)}$ completes the proof.   ∎

Lemma 2.3 introduces an interesting form of interpolation matrix (2.24) which is different from the interpolation matrix (2.2). The proof of the lemma is important in the sense that it reveals that the interpolation matrix $A(Z_1, \ldots, Z_N)$ is nonsingular for all Z_i in a certain neighborhood of ξ_i. However, Lemma 2.3 does not give any estimate on the sizes of ε or θ.

In [123, Theorem 3.5], by choosing ε to satisfy Lemma 2.3 and

$$0 < \varepsilon < h := \frac{1}{2} \min_{i \neq j} \|\xi_i - \xi_j\| \quad (2.26)$$

it is shown that the regressor vector $S(Z(t))$ (2.22) is persistently exciting if $Z(t)$ satisfies a kind of ergodic condition. This theorem is stated as follows.

PROPOSITION 2.1
Let I be a bounded μ-measurable subset of $[0, \infty)$ (take $I = [t_0, t_0 + T_0]$), and let sets I_i be given by

$$I_i = \{t \in I : \|Z(t) - \xi_i\| \leq \varepsilon\}, \qquad i = 1, \ldots, N \tag{2.27}$$

where ε is as in Lemma 2.3, subject to restriction (2.26).
For every $t_0 > 0$ and $T_0 > 0$, if $\mu(I_i) \geq \tau_0$ $(i = 1, \ldots, N)$, where τ_0 is a positive constant independent of t_0, then $S(Z(t))$ is persistently exciting in the sense of (2.16).

PROOF With the restriction (2.26), the balls with centers ξ_i and radius ε are nonintersecting, so that the subsets I_i given by (2.27) are disjoint, and consequently, the following inequality (Equation 3.4 in [123]) holds for every constant vector $c \in R^N$,

$$\int_I |S(Z)^T c|^2 d\mu(\tau) \geq \sum_{i=1}^N \int_{I_i} |S(Z)^T c|^2 d\mu(\tau) \tag{2.28}$$

Since

$$|S(Z(t))^T c|^2 = \left| \sum_{j=1}^N s(\|Z(t) - \xi_j\|)c_j \right|^2 \tag{2.29}$$

and $t \in I_i$ implies that $\|Z(t) - \xi_i\| \leq \varepsilon$, the following inequality is obtained

$$\max_{\|Z(t)-\xi_i\|\leq\varepsilon} \left\{ \left| \sum_{j=1}^N s(\|Z - \xi_j\|)c_j \right|^2 \right\} \int_{I_i} d\mu(\tau)$$

$$\geq \int_{I_i} |S(Z(t))^T c|^2 d\mu(\tau) \geq$$

$$\min_{\|Z(t)-\xi_i\|\leq\varepsilon} \left\{ \left| \sum_{j=1}^N s(\|Z - \xi_j\|)c_j \right|^2 \right\} \int_{I_i} d\mu(\tau) \tag{2.30}$$

where the maximum and minimum are taken over the ball $\|Z(t)-\xi_i\| \leq \varepsilon$ $(j = 1, \ldots, N)$. Due to the continuity of $|\sum_{j=1}^N s(\|Z - \xi_j\|)c_j|^2$ over this compact and connected ball, by using the Intermediate Value Theorem (see [110]), it is deduced that there exist $Z_i \in R^q$ such that $\|Z_i - \xi_i\| \leq \varepsilon$ and

$$\int_{I_i} |S(Z(t))^T c|^2 d\mu(\tau) = \left| \sum_{j=1}^N s(\|Z_i - \xi_j\|)c_j \right|^2 \mu(I_i) \tag{2.31}$$

holds for the nonintersecting subset I_i.

With $\mu(I_i) \geq \tau_0$ for $i = 1, \ldots, N$, we have

$$\int_I |S(Z(\tau))^T c|^2 d\mu(\tau) \geq \sum_{i=1}^N \left| \sum_{j=1}^N s(\|Z_i - \xi_j\|) c_j \right|^2 \tau_0$$

$$= \|Ac\|^2 \tau_0$$

holds for every constant vector $c \in R^N$, where A is the $N \times N$ matrix given by Equation (2.24). Because inequality $\|Ac\|^2 \geq \theta^2 \|c\|^2$ holds according to Lemma 2.3, the following inequality is obtained:

$$\int_{t_0}^{t_0+T_0} |S(Z(\tau))^T c|^2 d\mu(\tau) \geq \alpha_1 \|c\|^2, \qquad \alpha_1 = \theta^2 \tau_0$$

On the other hand, Lemma 2.2 implies that

$$\int_{t_0}^{t_0+T_0} |S(Z(\tau))^T c|^2 d\mu(\tau) \leq \alpha_2 \|c\|^2, \qquad \alpha_2 = s(O)^2 N T_0$$

Since both α_1 and α_2 are independent of t_0, it is therefore concluded that $S(Z(t))$ is persistently exciting in the sense of (2.16). ■

Proposition 2.1 states that for the regressor vector $S(Z(t))$ to be persistently exciting, the orbit $Z(t)$ must be ergodic in the sense that it visits in each time interval $[t_0, t_0 + T_0]$ a sufficiently small ε-ball about each neuron center ξ_i for a minimum amount of time that is independent of t_0. A simple example is a periodic orbit $Z(t)$ with period T_0 visiting the small ε-neighborhood of each neuron center for a minimum amount of time $\tau_0 > 0$ [123]. However, there are two related issues that need to be further addressed:

1. With the restriction on ε by Lemma 2.3 and Equation (2.26), it is possible that a particular periodic orbit does not visit the specified neighborhood of many neuron centers of the RBF network. Thus, Proposition 2.1 may not be applicable to practical RBF network identification and control of nonlinear systems.

2. To make the result applicable in practice, it is required to extend the restrictions on ε such that any periodic orbit will yield a regressor subvector consisting of every nearby neuron center. Note that in Lemma 2.3, the size of ε is not analyzed, only the existence of an $\varepsilon > 0$ is obtained. It is clear that when the restriction on ε is larger, Lemma 2.3, as well as Proposition 2.1, may not be valid.

To make Proposition 2.1 applicable to practical NN identification and control, it is necessary to remove the restrictions on ε, so that almost any periodic or periodic-like trajectory $Z(t)$ can lead to the satisfaction of the PE condition. As mentioned above, the restriction (2.26) was made to guarantee that the balls with centers ξ_i and radius ε are non-intersecting so that inequality (2.28) holds. This restriction, however, is actually unnecessary and can be enlarged.

For the regular lattice upon which the RBF network (2.8) is constructed, we choose

$$\varepsilon \geq \sqrt{q}h = \frac{\sqrt{q}}{2} \min_{i \neq j} \|\xi_i - \xi_j\| > 0 \tag{2.32}$$

Then, a periodic trajectory $Z(t)$ staying within the regular lattice will always yield a regression subvector $S_\zeta(Z)$ consisting of RBFs centered in an ε-neighborhood of the periodic trajectory $Z(t)$

$$S_\zeta(Z) = [s(\|Z_1 - \xi_{j_1}\|), \ldots, s(\|Z_{N_\zeta} - \xi_{j_{N_\zeta}}\|)]^T \in R^{N_\zeta} \tag{2.33}$$

where $\xi_{j_1}, \ldots, \xi_{j_{N_\zeta}}$ are distinctive centers. Moreover, since radial basis functions $s(\cdot)$ decay quickly and are small far from the centers, it is reasonable to choose

$$\sqrt{q}h \leq \varepsilon \leq \varepsilon' \tag{2.34}$$

such that for all Z_i $(i = 1, \ldots, N_\zeta)$ satisfying $\|Z_i - \xi_{j_i}\| < \varepsilon'$ we have $|s(\|Z_i - \xi_{j_i}\|)| > \iota$ where ι is a small positive constant.

We present the following theorem characterizing the PE property of the regression subvector $S_\zeta(Z(t))$. This result is based on our papers [242,243] with further extensions.

THEOREM 2.7
Consider a periodic trajectory $Z(t)$ with period T_0. Assume that $Z(t)$ is a continuous map from $[0, \infty)$ into a compact set $\Omega \subset R^q$, and $\dot{Z}(t)$ is bounded within Ω. Then, for the localized RBF network $W^T S(Z)$ (2.8) with centers placed on a regular lattice (large enough to cover the compact set Ω), the regressor subvector $S_\zeta(Z(t))$ as defined by (2.33) and (2.34), is persistently exciting in the sense of (2.16) almost always.

PROOF The proof of the theorem is done in two parts to overcome the two aforementioned issues.

(i) Take $I = [t_0, t_0 + T_0]$. Define subsets I_i in the same way as (2.27):

$$I_i = \{t \in I: \|Z(t) - \xi_{j_i}\| \leq \varepsilon\}, \qquad i = 1, \ldots, N_\zeta \tag{2.35}$$

For an arbitrary periodic trajectory $Z(t)$ with period T_0, we have $\mu(I_i) > \tau_0$.

When $\varepsilon \geq \sqrt{q}h > 0$, it is true that the sets I_i given in Equation (2.35) may be overlapping. To solve this problem, our idea is to divide the time that the orbit $Z(t)$ stays within the intersecting balls. Specifically, we describe I_i (2.35) in the following form:

$$I_i = I_{i_0} + I_{i_1} + \cdots + I_{i_Q} \tag{2.36}$$

where $1 \leq Q \leq N_\zeta - 1$, I_{i_0} represents the time that orbit $Z(t)$ visits and only visits the ball centered at ξ_{j_i}, and I_{i_k} $(k = 1, \ldots, Q)$ is the subset of I_i, representing the time that orbit $Z(t)$ simultaneously visits and only visits k other intersecting balls.

Note that I_{i_k} $(k = 1, \ldots, Q)$ being non-empty means that the trajectory $Z(t)$ will simultaneously visit the ε neighborhoods of $k+1$ neuron centers. Denote

$$I_i' = I_{i_0} + \frac{1}{2}I_{i_1} + \cdots + \frac{1}{Q+1}I_{i_Q} \tag{2.37}$$

where $\frac{1}{k+1}I_{i_k}$ $(k = 1, \ldots, Q)$ represents the divided piece of time that trajectory $Z(t)$ visits the intersecting $k+1$ balls. Note that if $\mu(I_i) > \tau_0$, then $\mu(I_i') > \tau_0'$, with $\tau_0 \geq \tau_0' > 0$.

Thus, the jointed sets I_i are turned into non-intersecting sets I_i', from which we have

$$\int_I |S_\zeta(Z(\tau))^T c|^2 d\mu(\tau) = \sum_{i=1}^{N_\zeta} \int_{I_i'} |S_\zeta(Z(\tau))^T c|^2 d\mu(\tau) \tag{2.38}$$

holds for every constant vector $c = [c_{j_1}, \ldots, c_{j_{N_\zeta}}]^T \in R^{N_\zeta}$ (with a little abuse of notation).

(ii) As we study the PE property of $S_\zeta(Z) = [s(\|Z_1 - \xi_{j_1}\|), \ldots, s(\|Z_{N_\zeta} - \xi_{j_{N_\zeta}}\|)]^T \in R^{N_\zeta}$, it is necessary to investigate the nonsingularity property of the following interpolation matrix:

$$A_\zeta = A_\zeta(Z_1, \ldots, Z_{N_\zeta})$$

$$= \begin{bmatrix} s(\|Z_1 - \xi_{j_1}\|) & \cdots & s(\|Z_1 - \xi_{j_{N_\zeta}}\|) \\ \vdots & \ddots & \vdots \\ s(\|Z_{N_\zeta} - \xi_{j_1}\|) & \cdots & s(\|Z_{N_\zeta} - \xi_{j_{N_\zeta}}\|) \end{bmatrix} \tag{2.39}$$

It is clear that when ε is given by Equation (2.34), A_ζ is not always nonsingular for all Z_i satisfying $\|Z_i - \xi_{j_i}\| \leq \varepsilon$. Thus, we need to investigate the following question: in the case when the interpolation matrix $A_\zeta(Z_1, \ldots, Z_{N_\zeta})$ is singular for some $Z = [Z_1, \ldots, Z_{N_\zeta}]^T$, does there always exist a $Z_0 = [Z_{10}, \ldots, Z_{N_\zeta 0}]^T$ in the neighborhood of Z (and also satisfying $\|Z_{i0} - \xi_{j_i}\| \leq \varepsilon$), such that $A_\zeta(Z_{10}, \ldots, Z_{N_\zeta 0})$ is nonsingular?

The answer to the question is given as follows. Because $\det(A_\zeta(Z))$ is a composite function of radial basis functions $(\cdot)$, it is an analytic function of $Z_1, \ldots, Z_{N_\zeta}$ (see, e.g., [255]). According to Lemma 2.3, $\det(A_\zeta(Z))$ is not identically zero, which means that the analytic function $\det(A_\zeta(Z))$ is generically non-zero; that is,

$$V_Z = \{Z| \det(A_\zeta(Z)) \neq 0\} \tag{2.40}$$

is open and dense. Equivalently,

$$V_Z^c = \{Z| \det(A_\zeta(Z)) = 0\} \tag{2.41}$$

is nowhere dense.

Thus, if Z_i ($i = 1, \ldots, N_\zeta$) are such that $\det(A_\zeta(Z)) = 0$ and Ω_0 is an open neighborhood of Z, then there always exists $Z_0 \in \Omega_0$, such that $\det(A_\zeta(Z_0)) \neq 0$ which means that $A_\zeta(Z_{10}, \ldots, Z_{N_\zeta 0})$ is nonsingular.

Moreover, as shown in the proof of Lemma 2.3, because $\lambda(Z)$ is a continuous function of Z_i ($i = 1, \ldots, N_\zeta$), there still exist $\theta' > 0$ such that

$$\|A_\zeta(Z_{10}, \ldots, Z_{N_\zeta 0})c\| \geq \theta'\|c\| \tag{2.42}$$

holds for all $c \in R^{N_\zeta}$.

Note that although the set (2.41) is nowhere dense, it may still form certain kinds of periodic trajectories. On the other hand, the open and dense set (2.40) implies that almost every periodic trajectory $Z(t)$ (except those described by Equation [2.41]) can ensure that (2.42) holds and the PE condition is satisfied. Specifically, we define $I_i'' \subseteq I_i'$ as the largest connected subset satisfying (2.40). It is clear that

$$\int_{I_i'} |S_\zeta(Z(t))^T c|^2 d\mu(\tau) \geq \int_{I_i''} |S_\zeta(Z(T))^T c|^2 d\mu(\tau)$$

From Equation (2.38), because

$$|S_\zeta(Z(t))^T c|^2 = \left| \sum_{j_i=j_1}^{j_{N_\zeta}} s(\|Z(t) - \xi_{j_i}\|)c_{j_i} \right|^2 \tag{2.43}$$

and $t \in I_i'$ still implies that $\|Z(t) - \xi_{j_i}\| \leq \varepsilon$, we still have the following inequality:

$$\max_{t \in I_i''} \left\{ \left| \sum_{j_i=j_1}^{j_{N_\zeta}} s_\zeta(\|Z(t) - \xi_{j_i}\|)c_{j_i} \right|^2 \right\} \int_{I_i'} d\mu(\tau)$$

$$\geq \int_{I_i'} |S_\zeta(Z(t))^T c|^2 d\mu(\tau)$$

$$\geq \min_{t \in I_i''} \left\{ \left| \sum_{j_i=j_1}^{j_{N_\zeta}} s(\|Z(t) - \xi_{j_i}\|)c_{j_i} \right|^2 \right\} \int_{I_i'} d\mu(\tau) \tag{2.44}$$

holds for all $Z(t)$ within the compact and connected region $\Omega' = \{z | \|z(t) - \xi_{ji}\| \leq \varepsilon, t \in I_i''\}$. By using the Intermediate Value Theorem (see [110]), there exist $Z_i \in \Omega'$ ($i = 1, \ldots, N_\zeta$) such that

$$\int_{I_i''} |S_\zeta(Z(t))^T c|^2 d\mu(\tau) = \left| \sum_{j_i=j_1}^{j_{N_\zeta}} s(\|Z_i - \xi_{j_i}\|)c_{j_i} \right|^2 \mu(I_i'') \tag{2.45}$$

holds for the non-intersecting subset I_i''.

With $\mu(I_i'') \geq \tau_0''$ for $i = 1, \ldots, N_\zeta (\tau_0 \geq \tau_0' \geq \tau_0'' > 0)$, we have

$$\int_I |S_\zeta(Z(t))^T c|^2 d\mu(\tau) \geq \sum_{i=1}^{N_\zeta} \int_{I_i''} |S_\zeta(Z(\tau))^T c|^2 d\mu(\tau)$$

$$\geq \sum_{i=1}^{N_\zeta} \left| \sum_{j_i=j_1}^{j_{N_\zeta}} s(\|Z_i - \xi_{j_i}\|) c_{j_i} \right|^2 \tau_0''$$

$$= \|A_\zeta c\|^2 \tau_0' \geq \theta'^2 \tau_0'' \|c\|^2$$

$$= \alpha_1' \|c\|^2, \quad \alpha_1' = \theta'^2 \tau_0''$$

holds for every constant vector $c \in R^{N_\zeta}$.

Therefore, similar to the other steps in the proof of Proposition 2.1, we have that for every constant vector $c \in R^{N_\zeta}$

$$\alpha_1' \|c\|^2 \leq \int_{t_0}^{t_0+T_0} |S_\zeta(Z(t))^T c|^2 d\mu(\tau) \leq \alpha_2 \|c\|^2$$

which means that for almost any periodic trajectory $Z(t)$, the corresponding regressor subvector $S_\zeta(Z(t))$ consisting of RBFs centered within the ε-neighborhood of the trajectory $Z(t)$ is persistently exciting. This ends the proof. ∎

REMARK 2.4

In the literature, satisfying the PE condition *a priori* has been considered as a difficult problem for identification and control of nonlinear systems. From the above analysis, we show that almost any periodic orbit can lead to the satisfaction of the (partial) PE condition. The significance of this result lies in that, with the partial PE condition satisfied, locally accurate NN approximation of unknown system dynamics can be achieved in identification and adaptive control of nonlinear systems using localized RBF networks.

What is shown in the above proof is that for almost any bounded trajectory $Z(t)$, as long as it stays within the regular lattice within which the RBF network is constructed, and passes through certain neurons centered within a neighborhood of trajectory $Z(t)$ at least once in a finite period of time, it will lead to the satisfaction of PE of a corresponding regressor subvector $S_\zeta(Z)$. This is actually the property of the class of recurrent trajectories in dynamical systems theory [206]. A recurrent trajectory represents a large set of periodic and periodic-like trajectories generated from nonlinear dynamical systems. Roughly speaking, a recurrent trajectory is characterized as: given $\xi > 0$, there exists $T(\xi) > 0$, such that the trajectory returns to the ξ-neighborhood of any point on the trajectory within a time not greater than $T(\xi)$. A remarkable feature of a recurrent trajectory is that regardless of the choice of the initial condition, given ξ, the whole trajectory lies in the ξ-neighborhood of the segment of the trajectory corresponding to a time interval $T(\xi)$ which is bounded [206]. Note that in contrast to periodic trajectories, whose return times are fixed, the return time for a recurrent trajectory is not fixed but is finite.

Recurrent trajectories frequently arise from nonlinear dynamical systems, including not only periodic trajectories, but also quasi-periodic, almost-periodic, and even some chaotic trajectories [206]. The following result is to establish a relationship between the recurrent trajectory and the PE condition, that is, to characterize the partial PE property of the corresponding regressor subvector for a recurrent trajectory.

COROLLARY 2.1
Consider a recurrent trajectory $Z(t)$ with "period" $T(\xi)$ in the sense defined above. Assume that $Z(t)$ is a continuous map from $[0, \infty)$ into a compact set $\Omega \subset R^q$, and $\dot{Z}(t)$ is bounded within Ω. Then, for the localized RBF network $W^T S(Z)$ (2.8) with centers placed on a regular lattice (large enough to cover the compact set Ω), the regressor subvector $S_\zeta(Z(t))$ as defined by (2.33) and (2.34), is persistently exciting in the sense of (2.16) almost always.

PROOF For a recurrent trajectory $Z(t)$ as described above, the whole trajectory lies in the ξ-neighborhood of a segment of the recurrent trajectory corresponding to a time interval $T(\xi)$ which is bounded.

Consider the regressor subvector $S_\zeta(Z(t))$ (as defined in Equation [2.33]), which consists of RBF neurons centered within an ε-neighborhood of the segment of the trajectory $Z(t)$. Then, the whole trajectory $Z(t)$ will visit an $(\varepsilon + \xi)$-neighborhood of those neurons on each time interval $[t_0, t_0 + T(\xi)]$ for a minimum amount of time. Because it is the nonsingularity property of the corresponding interpolation matrix that plays an important role, by following the other steps in the proofs of Theorem 2.7 and [123, Theorem 3.5], it is concluded that for almost any recurrent trajectory $Z(t)$, a corresponding regressor subvector $S_\zeta(Z(t))$ of the trajectory $Z(t)$ is persistently exciting. ∎

REMARK 2.5
The essential feature distinguishing periodic, quasi-periodic, and almost-periodic trajectories from recurrent chaotic trajectories lies in that, although the former ones have the property of uniform stability in the sense of Lyapunov, a recurrent chaotic trajectory is Lyapunov unstable (see [206] for more discussions). The instability of recurrent chaotic trajectories leads to the properties of divergence of nearby trajectories and sensitivity to initial conditions. Such properties yield the long-term unpredictable behaviors of nonlinear chaotic systems. In the above, it is shown that the satisfaction of the partial

PE condition does not require the trajectory $Z(t)$ to be Lyapunov stable. Thus, even an unpredictable chaotic trajectory, as long as it is recurrent, can satisfy the partial PE condition.

2.4   Summary

In this chapter, basic results about RBF network approximation and persistence of excitation have been presented. The main result is an improved characterization of the PE property of localized RBF networks driven by a periodic or recurrent trajectory.

With the partial PE condition satisfied for recurrent trajectories, we will show in the following chapters that the system dynamics of nonlinear dynamical systems undergoing recurrent motions (including the complex chaotic motions) can be accurately identified.

3

The Deterministic Learning Mechanism

In this chapter, we study the fundamental problem of how to achieve learning (i.e., knowledge acquisition) from unknown dynamical environments using neural networks (NN). This problem is related to system identification, the objective of which is to build mathematical models for dynamical systems based on observed data from the system. In system identification, the two mainstream approaches that dominate the field are subspace identification (see, e.g., [105]) and prediction error identification (see, e.g., [140]). Although the two approaches have been successful in identification of single-input single-output (SISO) and multi-input multi-output (MIMO) linear systems, identification of nonlinear dynamical systems still needs further research.

In identification of nonlinear dynamical systems, the neural network paradigm has been used due to its power for learning complex input-output mappings [162]. Since the 1990s, design and analysis of NN identification algorithms based on Lyapunov stability theory has attracted considerable interest from the adaptive control community [114,115,143,179]. Lyapunov-based identification is very attractive because it can provide a general formulation for modeling, identifying, and controlling nonlinear dynamical systems using neural networks. Analytical results concerning the stability of all the signals in the closed-loop system can be obtained, and convergence of the state estimation error to a small neighborhood of zero can be achieved. However, accurate estimation of system states does not necessarily lead to an accurate modeling or identification for system dynamics. In other words, the NN weight estimates normally are not guaranteed to converge to their optimal values. Without an effective identification for system dynamics, this kind of Lyapunov-based NN identification (via state estimation) may be useless, because nothing useful can be learned by the neural networks, and no constant information can be stored and reused for further recognition of the same or similar dynamical systems and their dynamical behaviors.

In this chapter, we investigate the problem of identification of nonlinear dynamical systems undergoing periodic or periodic-like motions. We have shown in the preceding chapter that the localized RBF network has the desired properties of function approximation and especially of satisfaction of a partial PE condition for periodic or periodic-like orbits. With the partial PE condition satisfied, by using a dynamical version of the localized radius basis function (RBF) network and a Lyapunov-based adaptation law for the RBF neural weights, the identification error system consisting of the state estimation error subsystem and weight estimation error subsystem can be proved to be exponentially stable along the periodic or periodic-like orbit. For neurons whose centers are close to the orbit, the neural weights will converge to small neighborhoods of a set of optimal values; whereas for the other neurons with centers far away from the orbit, the neural weights are not activated and are almost unchanged. Thus, accurate identification of the unknown dynamics can be achieved within a local region along the recurrent orbit. This means that a partial true system model can be accurately identified.

We refer to the above Lyapunov-based NN identification with the *a priori* verified partial PE condition as the deterministic learning mechanism. A comparison of the deterministic learning mechanism with conventional results of system identification is included in Section 3.4. Based on the deterministic learning mechanism, a learning theory is developed in the following chapters to constitute a new deterministic framework for knowledge acquisition, representation, and knowledge utilization in dynamical environments.

The results presented in this chapter are based on the authors' papers [238,244].

3.1   Problem Formulation

Consider a general nonlinear dynamical system in the following form:

$$\dot{x} = F(x; p), \qquad x(t_0) = x_0 \tag{3.1}$$

where $x = [x_1, \ldots, x_n]^T \in R^n$ is the state of the system, which is measurable, p is a constant vector of system parameters (different p will in general produce different dynamical behaviors), and $F(x; p) = [f_1(x; p), \ldots, f_n(x; p)]^T$ is a smooth but unknown nonlinear vector field.

ASSUMPTION 3.1

Assume that the state x remains uniformly bounded; that is, $x(t) \in \Omega \subset R^n$, $\forall t \geq t_0$, where Ω is a compact set. Moreover, the system trajectory starting from x_0, denoted as $\varphi_\zeta(x_0)$, is in either a periodic or periodic-like (recurrent) motion.

The following dynamical model using the RBF network is employed:

$$\dot{\hat{x}} = -A(\hat{x} - x) + \widehat{W}^T S_A(x) \tag{3.2}$$

where $\hat{x} = [\hat{x}_1, \ldots, \hat{x}_n]^T$ is the state vector of the dynamical model, x is the state of system (3.1), $A = diag\{a_1, \ldots, a_n\}$ is a diagonal matrix, with $a_i > 0$ being design constants, localized RBF networks $\widehat{W}^T S_A(x) = [\widehat{W}_1^T S_1(x), \ldots, \widehat{W}_n^T S_n(x)]^T$ are used to approximate the unknown $F(x; p) = [f_1(x; p), \ldots, f_n(x; p)]^T$ in Equation (3.1) within the compact set Ω, with each RBF network $\widehat{W}_i^T S_i(x)$ given by Equation (2.8) and $S_A(x) = diag\{S_1(x), \ldots, S_n(x)\}$.

The problem is to identify the unknown system dynamics $F(x; p)$ using only the information of system state $x(t)$. Specifically, the objective is to develop an adaptive NN identifier

$$\dot{\widehat{W}} = H(x, \hat{x}, \widehat{W}, t) \tag{3.3}$$

such that along the trajectory $\varphi_\zeta(x_0)$, a locally-accurate approximation of the unknown vector field $F(x; p)$ can be obtained by the RBF network $\widehat{W}^T S(x)$ and $\overline{W}^T S(x)$, where $\overline{W}$ is a constant vector obtained from $\widehat{W}$ according to some averaging procedure.

REMARK 3.1
It can be seen that the objective is not so ambitious, in the sense that accurate identification of $F(x, p)$ is not to be achieved in the whole space of interest, but only in a local region along the periodic or periodic-like system trajectory.

In the literature of Lyapunov-based identification, convergence of the state estimation error $\tilde{x} = \hat{x} - x$ to a small neighborhood of zero and the boundedness of the NN weight estimates $\widehat{W}$ can be achieved. However, convergence of the NN weight estimates $\widehat{W}$ to the optimal values W^* and accurate identification of system dynamics $F(x; p)$ normally cannot be achieved by $\widehat{W}^T S(x)$ unless a certain PE condition is satisfied. This actually implies that nothing can be learned in such an identification process without PE. Because the NN weight estimates $\widehat{W}$ are updated online and will continuously evolve according to the adaptation law (Equation [3.3]), the resulting $\widehat{W}$ are time-varying in nature. To identify $F(x; p)$, even a time-varying weight vector $\widehat{W}$ may be good enough for obtaining a sufficiently good approximation of the unknown system dynamics; the time-varying nature of $\widehat{W}$ (without converging to a constant vector) makes it very difficult to store and to reuse for further recognition tasks. Therefore, it is very important to ensure the convergence of $\widehat{W}$ to a constant vector $\overline{W}$.

3.2 Locally Accurate Identification of Systems Dynamics

In this section, we present a deterministic mechanism for learning (identifying) the unknown dynamics $F(x; p)$ in the nonlinear dynamical system (3.1).

One problem in using neural networks for identifying dynamical systems is that the existence of NN approximation errors and external noises may cause the estimates of neural weights to drift to infinity. This instability phenomenon, known as parameter drift in the robust adaptive control literature [92], can be dealt with by a Lyapunov-based design using robustification techniques (such as projection, deadzone, σ-modification) to keep the neural weights estimates ultimately bounded [92]. In the next subsection, we first consider an identification scheme using σ-modification, in which the stability of all the signals in the closed-loop identification system are guaranteed, and accurate learning is obtained. In Subsection 3.2.2, we show that even without any robustification technique, it is still possible to achieve accurate identification with the satisfaction of a partial PE condition.

3.2.1   Identification with σ-Modification

The dynamical RBF network (Equation [3.2]) constitutes the state estimation system, which has the same order as the identified system (3.1). From Equations (3.1) and (3.2), the derivative of the state estimation error $\tilde{x}_i = \hat{x}_i - x_i$ satisfies

$$\begin{aligned} \dot{\tilde{x}}_i &= -a_i \tilde{x}_i + \widehat{W}_i^T S_i(x) - f_i(x; p) \\ &= -a_i \tilde{x}_i + \widetilde{W}_i^T S_i(x) - \epsilon_i \end{aligned} \tag{3.4}$$

where $\widetilde{W}_i = \widehat{W}_i - W_i^*$, $\widehat{W}_i$ is the estimate of W_i^*, and $\epsilon_i = f_i(x; p) - W_i^{*T} S_i(x)$ is the ideal approximation error, as described in Chapter 2. The weight estimates $\widehat{W}_i$ are updated by the Lyapunov-based learning law:

$$\dot{\widehat{W}}_i = \dot{\widetilde{W}}_i = -\Gamma_i S_i(x) \tilde{x}_i - \sigma_i \Gamma_i \widehat{W}_i \tag{3.5}$$

where $\Gamma_i = \Gamma_i^T > 0$, and $\sigma_i > 0$ is a small value. The term $-\sigma_i \Gamma_i \widehat{W}_i$, which is referred to as the σ-modification technique [92], is used to keep the boundedness of $\widetilde{W}_i$ as well as $\widehat{W}_i$ in the case where it tends to drift to infinity due to the existence of the NN approximation ϵ_i.

The following theorem indicates that learning of the unknown $f_i(x; p)$ can be achieved along the recurrent trajectory $\varphi_\zeta(x_0)$.

THEOREM 3.1

Consider the adaptive system consisting of the nonlinear dynamical system (3.1), the dynamical RBF network (3.2), and the NN weight updating law (3.5). For almost any recurrent trajectory $\varphi_\zeta(x_0)$ starting from an initial condition $x_0 = x(0) \in \Omega$, and with initial values $\widehat{W}_i(0) = 0$, we have: (i) all signals in the adaptive system remain bounded; (ii) the state estimation error $\tilde{x}_i = \hat{x}_i(t) - x_i(t)$ converges exponentially to a small neighborhood around zero, and the neural-weight estimates $\widehat{W}_{\zeta i}$ (as given in [3.11]) converge to small neighborhoods of their optimal values $W_{\zeta i}^*$; and (iii) a locally accurate approximation for the unknown $f_i(x; p)$ to the desired error level ϵ_i is obtained along the trajectory $\varphi_\zeta(x_0)$ by either $\widehat{W}_i^T S_i(x)$ or $\overline{W}_i^T S_i(x)$ (as given in [3.15]).

PROOF   (i) For the adaptive system, consider the following Lyapunov function candidate:

$$V = \frac{1}{2} \tilde{x}_i^2 + \frac{1}{2} \widetilde{W}_i^T \Gamma_i^{-1} \widetilde{W}_i \tag{3.6}$$

The derivative of V along solutions of (3.4) is

$$\dot{V} = \tilde{x}_i \dot{\tilde{x}}_i + \widetilde{W}_i^T \Gamma_i^{-1} \dot{\widetilde{W}}_i$$
$$= -a_i \tilde{x}_i^2 - \tilde{x}_i \epsilon_i - \sigma_i \widetilde{W}_i^T \widehat{W}_i$$

Let $a_i = a_{i_1} + a_{i_2}$ with $a_{i_1}, a_{i_2} > 0$. Because

$$-a_{i_2} \tilde{x}_i^2 - \tilde{x}_i \epsilon_i \leq \frac{\epsilon_i^2}{4 a_{i_2}} \leq \frac{\epsilon_i^{*2}}{4 a_{i_2}}$$

$$-\sigma_i \widetilde{W}_i^T \widehat{W}_i \leq -\sigma_i \|\widetilde{W}_i\|^2 + \sigma_i \|\widetilde{W}_i\| \|W_i^*\| \leq -\frac{\sigma_i \|\widetilde{W}_i\|^2}{2} + \frac{\sigma_i \|W_i^*\|^2}{2}$$

it follows that

$$\dot{V} \leq -a_{i_1} \tilde{x}_i^2 - \frac{\sigma_i \|\widetilde{W}_i\|^2}{2} + \frac{\sigma_i \|W_i^*\|^2}{2} + \frac{\epsilon_i^{*2}}{4 a_{i_2}} \tag{3.7}$$

From the above, it is clear that $\dot{V}$ is negative definite whenever

$$|\tilde{x}_i| > \frac{\epsilon_i^*}{2\sqrt{a_{i_1} a_{i_2}}} + \sqrt{\frac{\sigma_i}{2 a_{i_1}}} \|W_i^*\| \quad \text{or} \quad \|\widetilde{W}_i\| > \frac{\epsilon_i^*}{\sqrt{2\sigma_i a_{i_2}}} + \|W_i^*\|.$$

This leads to the ultimate uniform boundedness of both $\tilde{x}_i$ and $\widetilde{W}_i$ as

$$|\tilde{x}_i| \leq \frac{\epsilon_i^*}{2\sqrt{a_{i_1} a_{i_2}}} + \sqrt{\frac{\sigma}{2 a_{i_1}}} \|W_i^*\| \tag{3.8}$$

$$\|\widetilde{W}_i\| \leq \frac{\epsilon_i^*}{\sqrt{2\sigma_i a_{i_2}}} + \|W_i^*\| \tag{3.9}$$

From the boundedness of x_i and W_i^*, we see that both $\hat{x}_i$ and $\widehat{W}_i$ are ultimately uniformly bounded. Thus, all the signals in the closed-loop system remain bounded. It is seen from Equation (3.8) that although $\tilde{x}_i$ can be made arbitrarily small with a_i large enough, no convergence result of $\|\widetilde{W}_i\|$ can be concluded from Equation (3.9), no matter how the design parameters are chosen.

(ii) Equations (3.4) and (3.5) constitute an adaptive system described in the following form:

$$\begin{bmatrix} \dot{\tilde{x}}_i \\ \dot{\widetilde{W}}_i \end{bmatrix} = \begin{bmatrix} -a_i & S_i(x)^T \\ -\Gamma_i S_i(x) & 0 \end{bmatrix} \begin{bmatrix} \tilde{x}_i \\ \widetilde{W}_i \end{bmatrix} + \begin{bmatrix} -\epsilon_i \\ -\sigma_i \widehat{W}_i \end{bmatrix} \tag{3.10}$$

According to Theorem 2.4, for the adaptive systems (3.10), when $S_i(x(t))$ is PE, the equilibrium point $(\tilde{x}_i, \widetilde{W}_i) = 0$ of the nominal part of system (3.10) is exponentially stable. However, PE of $S_i(x)$ requires the state $x(t)$ to visit every center of the whole RBF network "persistently," which is generally not feasible in practice.

By using the localization property of RBF networks, as shown in Equation (2.12), Equation (3.4) can be expressed in the following form along the trajectory $\varphi_\zeta(x_0)$:

$$\dot{\tilde{x}}_i = -a_i \tilde{x}_i + \widehat{W}_{\zeta i}^T S_{\zeta i}(x) + \widehat{W}_{\bar{\zeta} i}^T S_{\bar{\zeta} i}(x) - f_i(x); p)$$
$$= -a_i \tilde{x}_i + \widehat{W}_{\zeta i}^T S_{\zeta i}(x) - \epsilon'_{\zeta i} \qquad (3.11)$$

in which $(\cdot)_{\zeta i}$ and $(\cdot)_{\bar{\zeta} i}$ stand for terms related to the regions close to and away from the trajectory $\varphi_\zeta(x_0)$, respectively; $S_{\zeta i}(\chi)$ is a subvector of $S_i(x)$ as defined in Section 2.1; $\widehat{W}_{\zeta i}$ is the corresponding weight subvector; and $\epsilon'_{\zeta i} = \epsilon_{\zeta i} + \widehat{W}_{\bar{\zeta} i} S_{\bar{\zeta} i}(\chi) = 0(\epsilon_{\zeta i})$ is the approximation error along the trajectory $\varphi_\zeta(x_0)$.

The adaptive system (3.10) is now described by

$$\begin{bmatrix} \dot{\tilde{x}}_i \\ \dot{\tilde{W}}_{\zeta i} \end{bmatrix} = \begin{bmatrix} -a_i & S_{\zeta i}(x)^T \\ -\Gamma_{\zeta i} S_{\zeta i}(x) & 0 \end{bmatrix} \begin{bmatrix} \tilde{x}_i \\ \tilde{W}_{\zeta i} \end{bmatrix} + \begin{bmatrix} -\epsilon'_{\zeta i} \\ -\sigma_i \Gamma_{\zeta i} \widehat{W}_{\zeta i} \end{bmatrix} \qquad (3.12)$$

and

$$\dot{\widehat{W}}_{\bar{\zeta} i} = \dot{\tilde{W}}_{\bar{\zeta} i} = -\Gamma_{\bar{\zeta} i} S_{\bar{\zeta} i}(x) \tilde{x}_i - \sigma_i \Gamma_{\bar{\zeta} i} \widehat{W}_{\bar{\zeta} i} \qquad (3.13)$$

Based on the properties of RBF networks (as stated in Section 2.3), almost any periodic or recurrent trajectory $\varphi_\zeta(x_0)$ ensures PE of the regressor subvector $S_{\zeta i}(x)$. According to Theorem 2.4, when $S_{\zeta i}(x)$ is PE, the origin $(\tilde{x}_i, \tilde{W}_{\zeta i}) = 0$ of the nominal part of system (3.12) is exponentially stable. Because $\epsilon'_{\zeta i} = O(\epsilon_{\zeta i}) = O(\epsilon_i)$, and $\sigma \Gamma_{\zeta i} \widehat{W}_{\zeta i}$ can be made small by choosing σ small enough, by using Theorem 2.6, both the state error $\tilde{x}_i(t)$ and the parameter error $\tilde{W}_{\zeta i}(t)$ converge exponentially to some small neighborhoods of zero, with the sizes of the neighborhoods being determined, respectively, by ϵ_i^* and $\sigma_i \|\Gamma_{\zeta i} W_{\zeta i}^*\|$.

(iii) The convergence of $\widehat{W}_{\zeta i}$ to be in a small neighborhood of $W_{\zeta i}^*$ implies that along the trajectory $\varphi_\zeta(x_0)$,

$$f_i(\varphi_\zeta; p) = W_{\zeta i}^{*T} S_{\zeta i}(\varphi_\zeta) + \epsilon_{\zeta i} = \widehat{W}_{\zeta i}^T S_{\zeta i}(\varphi_\zeta) - \tilde{W}_{\zeta i}^T S_{\zeta i}(\varphi_\zeta) + \epsilon_{\zeta i}$$
$$= \widehat{W}_{\zeta i}^T S_{\zeta i}(\varphi_\zeta) + \epsilon_{\zeta i_1} \qquad (3.14)$$

where $\epsilon_{\zeta i_1} = \epsilon_{\zeta i} - \tilde{W}_{\zeta i}^T S_{\zeta i}(\varphi) = O(\epsilon_{\zeta i}) = O(\epsilon_i)$ is the practical approximation error for using $\widehat{W}_{\zeta i}^T S_{\zeta i}$, which is small due to the exponential convergence of $\tilde{W}_{\zeta i}$.

Again, by the convergence result, we can obtain a constant vector of neural weights according to

$$\overline{W}_i = \text{mean}_{t \in [t_a, t_b]} \widehat{W}_i(t) \qquad (3.15)$$

where "mean" is the arithmetic mean [39], and $t_b > t_a > 0$, represents a piece of time segment after the transient process. Thus, using $\overline{W}_{\zeta i}^T S_{\zeta i}(\varphi_\zeta)$, where

$\overline{W}_{\zeta i} = [\overline{w}_{j_1}, \ldots, \overline{w}_{j_\zeta}]^T$ is the subvector of $\overline{W}_i$, we have

$$f_i(\varphi_\zeta; p) = \widehat{W}_{\zeta i}^T S_{\zeta i}(\varphi_\zeta) + \epsilon_{\zeta i_1} = \overline{W}_{\zeta i}^T S_{\zeta i}(\varphi_\zeta) + \epsilon_{\zeta i_2} \tag{3.16}$$

where $\epsilon_{\zeta i_2}$ is the practical approximation error for using $\overline{W}_{\zeta i}^T S_{\zeta i}$. It is clear that after the transient process, $\epsilon_{\zeta i_2} = O(\epsilon_{\zeta i1}) = O(\epsilon_i)$. This implies that a certain part of the RBF network, represented by either $\widehat{W}_{\zeta i}^T S_{\zeta i}(x)$ or $\overline{W}_{\zeta i}^T S_{\zeta i}(x)$, is indeed capable of approximating the unknown nonlinearity $f_i(x; p)$ to the desired error level ϵ_i along the trajectory $\varphi_\zeta(x_0)$.

On the other hand, from the adaptation law (3.13), it can be seen that for the neurons with centers far away from the trajectory $\varphi_\zeta(x_0)$, $|S_{\bar{\zeta}}(\chi)|$ will become very small due to the localization property of RBFs. In this case, the neural weights $\widehat{W}_{\bar{\zeta}}$ will only be slightly updated. Both $\widehat{W}_{\bar{\zeta}}$ and $\widehat{W}_{\bar{\zeta} i}^T S_{\bar{\zeta} i}(x)$, as well as $\overline{W}_{\bar{\zeta} i}$ and $\overline{W}_{\bar{\zeta} i}^T S_{\bar{\zeta} i}(x)$, will remain very small. This means that the entire RBF network $W_i^T S_i(x)$ can approximate the unknown $f_i(x; p)$ along the trajectory $\varphi_\zeta(x_0)$ as following using Equation (3.14):

$$\begin{aligned} f_i(\varphi_\zeta; p) &= \widehat{W}_{\zeta i}^T S_{\zeta i}(\varphi_\zeta) + \epsilon_{\zeta i_1} \\ &= \widehat{W}_{\zeta i}^T S_{\zeta i}(\varphi_\zeta) + \widehat{W}_{\bar{\zeta} i}^T S_{\bar{\zeta} i}(\varphi_\zeta) + \epsilon_{\zeta i_1} - \widehat{W}_{\bar{\zeta} i}^T S_{\bar{\zeta} i}(\varphi_\zeta) \\ &= \widehat{W}_i^T S_i(\varphi_\zeta) + \epsilon_{i_1} \end{aligned} \tag{3.17}$$

where $\epsilon_{i_1} = \epsilon_{\zeta i_1} - \widehat{W}_{\bar{\zeta} i}^T S_{\bar{\zeta} i}(\varphi_\zeta) = O(\epsilon_{\zeta i1}) = O(\epsilon_i)$. Similarly, using Equation (3.14) we have

$$\begin{aligned} f_i(\varphi_\zeta; p) &= \overline{W}_{\zeta i}^T S_{\zeta i}(\varphi_\zeta) + \epsilon_{\zeta i_2} \\ &= \overline{W}_{\zeta i}^T S_{\zeta i}(\varphi_\zeta) + \overline{W}_{\bar{\zeta} i}^T S_{\bar{\zeta} i}(\varphi_\zeta) + \epsilon_{\zeta i_2} - \overline{W}_{\bar{\zeta} i}^T S_{\bar{\zeta} i}(\varphi_\zeta) \\ &= \overline{W}_i^T S_i(\varphi_\zeta) + \epsilon_{i_2} \end{aligned} \tag{3.18}$$

where $\epsilon_{i_2} = \epsilon_{\zeta i_2} - \overline{W}_{\bar{\zeta} i}^T S_{\bar{\zeta} i}(\varphi_\zeta) = O(\epsilon_{\zeta i_2}) = O(\epsilon_i)$. Equations (3.17) and (3.18) mean that locally accurate identification of the system dynamics to the desired error level ϵ_i can be achieved by using the entire RBF network, either $\widehat{W}_i^T S_i(x)$ or $\overline{W}_i^T S_i(x)$, in a local region along the trajectory. ∎

It is seen that the employment of localized RBF networks in Equation (3.2), under periodic or periodic-like (recurrent) inputs, yields a guaranteed, partial PE condition. This condition, with the localization property of RBF networks, leads to the exponential stability of a localized adaptive system. In this way, parameter convergence and accurate local identification of system dynamics take place naturally in the dynamical process.

REMARK 3.2
For the (possibly large) region where the trajectory does not explore, no learning occurs, as represented by the slightly updated $\widehat{W}_{\bar{\zeta} i}, \overline{W}_{\bar{\zeta} i}$, and small $\widehat{W}_{\bar{\zeta} i}^T S_{\bar{\zeta} i}(x)$ and $\overline{W}_{\bar{\zeta} i}^T S_{\bar{\zeta} i}(x)$. In fact, Equations (3.17) and (3.18) imply another advantage

obtained from the localization property of RBF networks. Accurate learning in a local region along the trajectory is achieved by using the entire RBF network $\widehat{W}_i^T S_i(x)$ or $\overline{W}_i^T S_i(x)$, as well as using the partial RBF network $\widehat{W}_{\zeta i}^T S_{\zeta i}(x)$ or $\overline{W}_{\zeta i}^T S_{\zeta i}(x)$. In other words, although useful knowledge is obtained only in $\widehat{W}_{\zeta i}$, it is not necessary to specify which neural weights belonging to $\widehat{W}_{\zeta i}$ need to be updated. For this reason, with the RBF network constructed on a regular lattice, we can update all the neural weights according to Equation (3.5), which makes the algorithm easily implementable.

REMARK 3.3

In the above, we did not give an explicit expression for the convergence rates of $\tilde{x}_i$ and $\widetilde{W}_i$. This requires the estimation of the excitation levels α_1 and α_2 in Equation (2.16) for RBF networks, and the establishment of a relationship between the PE condition and the exponential convergence rates, both of which are very complicated [123,199]. Nevertheless, it is possible for us to provide a brief discussion here on the parameter convergence rate, that is, the learning rate. As discussed in Section 2.3, we have

$$\alpha_1 \propto \tau_0, \qquad \alpha_2 \propto T \tag{3.19}$$

where τ_0 is the minimum amount of time that $Z(t)$ stays within a small neighborhood of the involved center, and T is the period by which the trajectory passes through each center of the RBF network. With PE of $S_{\zeta i}(\varphi_\zeta)$ being satisfied by system (3.12), a larger α_1 or a smaller α_2 will normally lead to a faster parameter convergence rate (see [199, Chapter 2]). Thus, it is concluded, and is verified by simulations, that a larger τ_0 and a smaller T will make the learning proceed at a faster rate. On the other hand, due to the existence of the NN approximation errors ϵ_i, it can be concluded from [110, Lemma 5.2] and [199, Chapter 2], that the actual parameter estimation errors (the learning error) are inversely proportional to α_1 and so to τ_0. Thus, a larger τ_0 will make learning more accurate.

3.2.2 Identification without Robustification

In the above, we used σ-modification [92] as one robustification technique to cope with the effect of NN approximation errors. Note that the boundedness results in step (i) of the above proof are obtained without the PE condition. The concern in this subsection is to investigate with a partial PE condition satisfied, whether it is possible to achieve accurate identification without using any robustification technique. In this case, the neural weights are updated by the following adaptation law:

$$\dot{\widehat{W}}_i = \dot{\widetilde{W}}_i = -\Gamma_i S_i(x)\tilde{x}_i \tag{3.20}$$

where the σ-modification term $-\sigma_i \Gamma_i \widehat{W}_i$ used in Equation (3.5) does not appear.

Previous analysis has shown that without robustification, the adaptation law (3.20) alone cannot guarantee the boundedness of $\widetilde{W}_i$ when $\tilde{x}_i$ becomes small. The existence of NN approximation errors ϵ_i may cause both $\widetilde{W}_i$ and $\widehat{W}_i$ to drift to infinity, a well-known instability phenomenon in robust adaptive control theory [92]. It is also shown that in the case where a complete PE condition of $S_i(x)$ is satisfied, it is not necessary to employ any robustification technique for the boundedness of the signals in the closed-loop system. However, what we have is not the PE of the entire regressor vector, but only a partial PE condition of a regressor subvector $S_{\zeta i}(x)$. The following corollary indicates that with this partial PE condition, accurate learning of the unknown dynamics $F(x; p)$ can still be achieved, even without robustification.

COROLLARY 3.1
Consider the adaptive system, consisting of the nonlinear dynamical system (3.1), the dynamical RBF network (3.2), and the NN weight updating law (3.20). For almost any recurrent trajectory $\varphi_\zeta(x_0)$ starting from initial condition $x_0 = x(0) \in \Omega$, and with initial values $\widehat{W}_i(0) = 0$, both the state estimation errors $\tilde{x}_i = \hat{x}_i(t) - x_i(t)$ and the NN weight estimation errors $\widetilde{W}_{\zeta i}$ converge exponentially to small neighborhoods around zero, and a locally accurate approximation of the unknown $f_i(x; p)$ to the desired error level ϵ_i is achieved along the recurrent trajectory $\varphi_\zeta(x_0)$.

PROOF Consider the following Lyapunov function:

$$V = \frac{1}{2}\sum_{i=1}^{n}(\tilde{x}_i^2 + \widetilde{W}_i^T \Gamma_i^{-1} \widetilde{W}_i) \tag{3.21}$$

By combining Equations (3.4) and (3.20), the derivative of V is

$$\dot{V} = \sum_{i=1}^{n}(\tilde{x}_i \dot{\tilde{x}}_i + \widetilde{W}_i^T \Gamma_i^{-1} \dot{\widehat{W}}_i)$$

$$= \sum_{i=1}^{n}(-a_i \tilde{x}_i^2 - \tilde{x}_i \epsilon_i)$$

$$\leq \sum_{i=1}^{n}\left(-\frac{1}{2}a_i \tilde{x}_i^2 + \frac{\epsilon_i^{*2}}{2a_i}\right) \tag{3.22}$$

It is clear that $\dot{V}$ is negative whenever $|\tilde{x}_i| > \frac{\epsilon_i^*}{a_i}$. This means that $\tilde{x}_i$ ($i = 1, \ldots, n$) will remain bounded for all time, and will eventually converge to a small neighborhood of zero bounded by $\frac{\epsilon_i^*}{a_i}$.

From the adaptation law (3.20), we have

$$\dot{\widehat{W}}_{\zeta i} = \dot{\widetilde{W}}_{\zeta i} = -\Gamma_{\zeta i} S_{\zeta i}(x)\tilde{x}_i \tag{3.23}$$

$$\dot{\widehat{W}}_{\bar{\zeta} i} = \dot{\widetilde{W}}_{\bar{\zeta} i} = -\Gamma_{\bar{\zeta} i} S_{\bar{\zeta} i}(x)\tilde{x}_i \tag{3.24}$$

With the boundedness of $\tilde{x}_i$, since $S_{\bar{\zeta} i}(x)$ is very small due to the localization property of RBF, it is concluded that each element of $\widehat{W}_{\bar{\zeta} i}$ will be kept small in a time interval $[t_0, T_0)$, where $T_0 > t_0$ could be very large.

Thus, within this time interval $[t_0, T_0)$, the state-estimation subsystem (3.4) can still be described by:

$$\dot{\tilde{x}}_i = -a_i \tilde{x}_i + \widehat{W}_{\zeta i}^T S_{\zeta i}(x) + \widehat{W}_{\bar{\zeta} i}^T S_{\bar{\zeta} i}(x) - f_i(x)$$

$$= -a_i \tilde{x}_i + \widetilde{W}_{\zeta i}^T S_{\zeta i}(x) - \epsilon_{\zeta i} + \widehat{W}_{\bar{\zeta} i}^T S_{\bar{\zeta} i}(x) \qquad (3.25)$$

The adaptive system (3.10) is now described by

$$\begin{bmatrix} \dot{\tilde{x}}_i \\ \dot{\widetilde{W}}_{\zeta i} \end{bmatrix} = \begin{bmatrix} -a_i & S_{\zeta i}(x)^T \\ -\Gamma_{\zeta i} S_{\zeta i}(x) & 0 \end{bmatrix} \begin{bmatrix} \tilde{x}_i \\ \widetilde{W}_{\zeta i} \end{bmatrix} + \begin{bmatrix} -\epsilon'_{\zeta i} \\ 0 \end{bmatrix} \qquad (3.26)$$

where $\epsilon'_{\zeta i} = \epsilon_{\zeta i} - \widehat{W}_{\bar{\zeta} i}^T S_{\bar{\zeta} i}(\varphi_\zeta) = O(\epsilon_{\zeta i}) = O(\epsilon_i)$. To this end, with the partial PE of the regressor subvector $S_{\zeta i}(x)$, exponential convergence of $(\tilde{x}_i, \widetilde{W}_{\zeta i})$ to small neighborhoods of zero can be achieved within the time interval $[t_0, T_0)$. The sizes of the neighborhoods are determined by $|\epsilon_i^*|$. This implies that $\widehat{W}_{\zeta i}$ will converge to small neighborhoods of their optimal values $W_{\zeta i}^*$, that is, converge to constant values, and $\widehat{W}_{\bar{\zeta} i}$ will remain small within the time interval $[t_0, T_0)$.

Therefore, using steps similar to those in Theorem 3.1, it can be concluded that within the time interval $[t_0, T_0)$, partial parameter convergence (deterministic learning) can be obtained, and locally accurate approximation of the unknown dynamics $f_i(x; p)$ to the desired error level ϵ_i can be achieved along the trajectory $\varphi_\zeta(x_0)$. ∎

REMARK 3.4

Compared to Theorem 3.1, the adaptive law (3.20) does not guarantee boundedness of all signals. However, it is seen that thanks to the properties of localization and partial PE of RBF networks, learning can take place within a finite time interval. Therefore, it is unnecessary to conduct stability analysis when time goes to infinity. Compared with the NN identification methods with robustification, an advantage without using robustification is that more accurate parameter convergence may be achieved, and improved approximation of system dynamics can be obtained.

3.3   Comparison with System Identification

In this section, we briefly discuss the connection between the deterministic learning mechanism and existing results on system identification.

System identification theory was developed around 1960 based on the introduction of the state-space representation by Kalman and Bartram [103] for model-based control design. In 1965, Astrom and Bohlin [9] introduced into the control community the ARMA (autoregressive moving average) or

ARMAX model (autoregressive moving average with exogenous inputs), which gave rise to the prediction error identification framework that has dominated identification theory and applications [10,11,139,212]. The objective of the prediction error method was to find conditions on the parameterization and the experimental conditions under which the estimated model would converge to the true system. For example, an input-output model structure is chosen as follows [136]:

$$y_t = G(z, \theta)u_t + H(z, \theta)e_t \tag{3.27}$$

where $G(z, \theta)$ and $H(z, \theta)$ are parameterized rational transfer functions and e_t is white noise. All commonly used prediction error model structures in linear system identification are special cases of the generic structure (3.27). Moreover, identification of nonlinear systems with known model structures but unknown parameters parallels the analysis and solution of linear identification problems. By introducing special classes of nonlinear black-box models such as Wiener, Hammerstein, splines, neural networks, and wavelets, a collective effort set a similar framework for identifying nonlinear black-box models [102,209].

To estimate θ from (3.27), the one-step-ahead prediction error is derived as $\varepsilon_t(\theta) = y_t - \hat{y}_{t|t-1}(\theta)$, where $\hat{y}_{t|t-1}(\theta)$ is the one-step-ahead prediction. Then, given a set of Z^N of N data, one can define an identification criterion as a nonnegative function of the prediction errors,

$$V_N(\theta, Z^N) = \frac{1}{N} \sum_{t=1}^{N} l(\varepsilon_t(\theta)) \tag{3.28}$$

where $l(\cdot)$ is a nonnegative scalar-valued function. Minimizing $V_N(\theta, Z^N)$ with respect to θ over a domain D_θ then yields the parameter estimate

$$\hat{\theta}_N = \text{argmin}_{\theta \in D_\theta} V_N(\theta, Z^N) \tag{3.29}$$

This is the well-known prediction error approach [139,212]. The convergence to the true parameters and the identification of the true system model relies on the satisfaction of the PE condition. It was soon realized that for linear system identification the PE condition can be satisfied only when the input u is informative enough or sufficiently rich in frequency domain. However, for nonlinear identification there is no relationship established between the frequencies of the input u and the parameters to be estimated. Consequently, the idea of identification of a true nonlinear system model was progressively abandoned [2,135,137].

When identification of a true system model is not the objective, identification then is considered as a design problem such that the estimated model is used for a specific purpose, for example, for the purpose of model-based control design, as control is often the main motivation for system identification in the systems and control community. Identification for control has also triggered new research activity on the identification of systems operating in closed loop [48]. In identification for model-based control, the basic idea is that as long as the control performance is achieved, the acceptance of the estimated models is justified by the "usefulness" rather than "truth" [70].

On the other hand, in conventional adaptive control (which also contains studies on system identification) [79,160], a significant question was apparently left incompletely resolved. Using state-space or the ARMAX-type input-output model, PE could be invoked to guarantee parameter convergence, and in this sense, accurate identification. However, the control task could be achieved without imposing PE. It was not clear what kind of PE is actually necessary for control.

From the above, it is seen that identification of a true nonlinear system model is too difficult to be achieved in conventional system identification, and identifying the true system model is considered unnecessary for model-based control. From our point of view, the difficulties for identification of a true nonlinear system model lie in the selection of the parametric model structure, which leads to the difficulty in satisfying the PE condition. With this difficulty, the problems of closed-loop identification and nonlinear system identification may not be easily resolved within the framework of the prediction error approach.

The deterministic learning mechanism presented in this chapter apparently provides a new viewpoint for system identification. By selecting the localized RBF networks as the parameterized model structure, parameters appear in a network in the form of the neural weights. When a periodic or periodic-like orbit is taken as the NN input, a direct connection is established between the periodic or periodic-like orbit and the estimated weights (parameters) of neurons centered in a local region along the periodic or periodic-like orbit. This leads naturally to the satisfaction of a partial PE condition. Consequently, exponential stability of the estimation systems is guaranteed, and convergence of neural weight estimates to small neighborhoods of their optimal values is obtained. Compared with existing system identification approaches, the main feature of the deterministic learning approach is that locally accurate identification of a partial true nonlinear system model is achieved in a local region along the periodic or periodic-like orbit. In this way, the problem of nonlinear system identification is partly resolved. Closed-loop identification of control system dynamics can be implemented in a similar way, as is described in Chapter 4. Furthermore, the obtained knowledge on identified partial system models can be stored and represented by constant RBF networks, and can be readily used for another similar control task toward guaranteed stability and improved control performance. For a number of model-based control tasks, the identifier produces a set of partial models or a multimodel that is connected to the tasks. Moreover, accurate identification of partial system models makes it possible to measure the similarity of control situations or dynamical patterns, to implement rapid recognition of dynamical patterns, and to establish the framework of pattern-based control. These aspects are dealt with in later chapters.

3.4 Numerical Experiments

To verify the deterministic learning approach described above, we take the following Rossler system [186] as an example:

$$\dot{x}_1 = -x_2 - x_3$$
$$\dot{x}_2 = x_1 + p_1 x_2$$
$$\dot{x}_3 = p_2 + x_3(x_1 - p_3) \qquad (3.30)$$

where $x = [x_1, x_2, x_3]^T \in R^3$ is the state vector which is available from measurement, $p = [p_1, p_2, p_3]^T$ is a constant vector of system parameters, and the system dynamics $f_1(x; p) = -x_2 - x_3$, $f_2(x; p) = x_1 + p_1 x_2$, and $f_3(x; p) = p_2 + x_3(x_1 - p_3)$ are assumed mostly unknown to the identifier. For convenience of presentation, we assume that the state variables of each function are known: for example, $f_2(x; p)$ is a function of (x_1, x_2), and $f_3(x; p)$ is a function of (x_1, x_3).

According to [28], by fixing $p_1 = p_2 = 0.2$, and varying p_3, the Rossler system (3.30) can generate a sequence of period-doubling bifurcations leading to chaos. For example, it exhibits a period-1 orbit when $p_3 = 2.5$ (Figure 3.1a), a period-2 orbit when $p_3 = 3.3$ (Figure 3.3 a), and a chaotic orbit when $p_3 = 4.5$ (Figure 3.5a).

The dynamical RBF networks (3.2) are used to identify the unknown system dynamics $f_i(x; p)$ $(i = 1, 2, 3)$ in Equation (3.30). We construct RBF network $\widehat{W}_i^T S_i(x)$ $(i = 1, 2, 3)$ with the centers μ_i evenly placed on $[-12, 12] \times [-2, 16]$, $[-12, 12] \times [-12, 12]$, and $[-12, 12] \times [-2, 16]$, respectively, and the widths $\eta_i = 0.5$ $(i = 1, \ldots, l)$. It is obvious that the three mentioned system orbits will not explore every center of the RBF networks.

The weights of the RBF networks are updated online according to Equation (3.20), that is, using the adaptation law without any robustification. The design parameters of the above controller are $c_i = 6$, $\Gamma_i = diag\{3, 3, 3\}$, $i = 1, 2, 3$. The initial weights are $\widehat{W}_i(0) = 0.0$, and the initial conditions $[x_1(0), x_2(0), x(3)]^T = [0.5, 0.2, 0.3]^T$ and $[\hat{x}_1(0), \hat{x}_2(0), \hat{x}_3(0)]^T = [0.2, 0.3, 0.0]^T$.

First, system (3.30) (with $p_3 = 2.5$) in period-1 orbit is to be identified. Figures 3.1a and b show the period-1 orbit in space and in the time domain. The convergence of the neural weights is shown in Figures 3.1c and d. Especially, Figure 3.1d demonstrates partial parameter convergence; that is, only the weight estimates of some neurons whose centers are close to the orbit are being activated and updated. These weight estimates converge to their optimal values $W_{\zeta i}^*$. Other neuron centers far away from the trajectory are not activated and updated, thus, their weight estimates, with the initial conditions being set to zero, are almost unchanged.

Since the optimal values $W_{\zeta i}^*$ are generally unknown, it is difficult to verify whether $\widehat{W}_{\zeta i}$ have indeed converged to $W_{\zeta i}^*$. Fortunately, we can show the NN

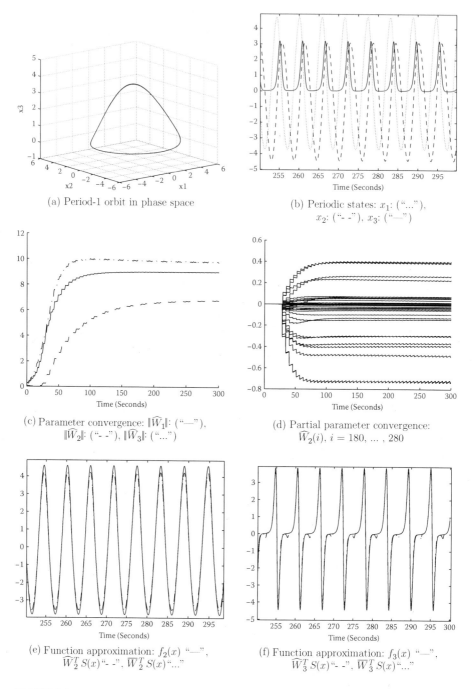

(a) Period-1 orbit in phase space

(b) Periodic states: x_1: ("..."),
x_2: ("- -"), x_3: ("—")

(c) Parameter convergence: $\|\widehat{W}_1\|$: ("—"),
$\|\widehat{W}_2\|$: ("- -"), $\|\widehat{W}_3\|$: ("...")

(d) Partial parameter convergence:
$\widehat{W}_2(i)$, $i = 180, \dots, 280$

(e) Function approximation: $f_2(x)$ "—",
$\widehat{W}_2^T S(x)$ "- -", $\overline{W}_2^T S(x)$ "..."

(f) Function approximation: $f_3(x)$ "—",
$\widehat{W}_3^T S(x)$ "- -", $\overline{W}_3^T S(x)$ "..."

FIGURE 3.1

Identification of the Rossler system (3.30) with a period-1 orbit ($p_3 = 2.5$).

approximations of $f_i(x; p)$ both in the time domain and in the phase space. For conciseness of presentation, only the NN approximations of the linear dynamics $f_2(x; p)$ and nonlinear dynamics $f_3(x; p)$ are presented in the sequel. From Figures 3.1e and f, we can see that good NN approximations of the unknown dynamics $f_2(x; p)$ and $f_3(x; p)$ are obtained. In Figures 3.2c and d, we show that accurate approximations of linear dynamics $f_2(x; p)$ and nonlinear dynamics $f_3(x; p)$ are indeed achieved along the period-1 orbit. Compared with the true system dynamics as shown in Figures 3.2a and b, the locally accurate NN approximations can be considered as partially true system dynamics $f_2(x; p)$ and $f_3(x; p)$ stored in constant RBF networks $\overline{W}_i^T S_i(x)$ ($i = 2, 3$), as shown in Figures 3.2e and f. For the space where the orbit does not explore, no learning occurs, as represented by the zero-plane in Figures 3.2e and f, due to the small values of $\overline{W}_i^T S_i(x)$ in that space.

Second, similar results are obtained in Figures 3.3 and 3.4 for identification of system (3.30) (with $p_3 = 3.3$) exhibiting a period-2 orbit. It is noticed from Figure 3.3c that the parameter convergence rates are slower as compared with those in Figure 3.1c. This is because, (i) as seen from Figures 3.1b and 3.3b, the period-2 trajectory has a larger T as compared with the period-1 trajectory; (ii) as the speed of the period-2 orbit appears to be faster in certain areas and more neurons are involved, the minimum amount of time τ for the orbit to pass through the neuron center in certain areas might be reduced. Thus, as discussed in Remark 3.3, in the period-2 case the excitation level α_1 becomes smaller and α_2 becomes larger. Consequently, the parameter convergence rates, as well as the rate of learning, are slower. However, it is seen from Figures 3.3 and 3.4 that good NN approximations can still be achieved within a longer time interval $[0, 800]$ seconds.

Third, we consider the chaotic orbit when $p_3 = 4.5$. As seen in Figure 3.5a, the chaotic orbit explores much larger areas in space, which implies that many more neurons are involved and activated, and the "period" T of the orbit becomes much larger compared with the above periodic orbits. The states $x_i(t)$ ($i = 1, 2, 3$) of the chaotic orbit are random-like signals, as shown in Figure 3.5a. Moreover, the state $x_3(t)$ has many spikes, which means that the speed of the chaotic orbit becomes much faster in certain areas in phase space.

According to Corollary 2.1, recurrent trajectories, including chaotic ones, can satisfy the PE condition. In Figure 3.5c, it is seen that within the time interval $[0, 800]$ seconds, $\|\widehat{W}_2\|$ nearly converges to a constant value. Partial parameter convergence is shown in Figure 3.5d. The NN approximation of linear dynamics $f_2(x; p)$ along the chaotic orbit is shown in Figures 3.5e and 3.6c and e. It can be seen from Figure 3.6c that with (x_1, x_2) as NN inputs to $W_2^T S_2(x)$, the linear dynamics $f_2(x; p)$ of the chaotic Rossler system can be accurately identified by both $\widehat{W}_2^T S_2(x)$ and $\overline{W}_2^T S_2(x)$.

In Figure 3.5c, it is also seen that within the time interval $t = [0, 800]$ seconds, $\|\widehat{W}_1\|$ and $\|\widehat{W}_3\|$ have not yet converged to constant values. This is mainly because both the NN inputs (x_2, x_3) to $\widehat{W}_1^T S_1(x)$ and (x_1, x_3) to $\widehat{W}_3^T S_3(x)$ include $x_3(t)$, which makes the trajectory move very quickly in certain areas in phase

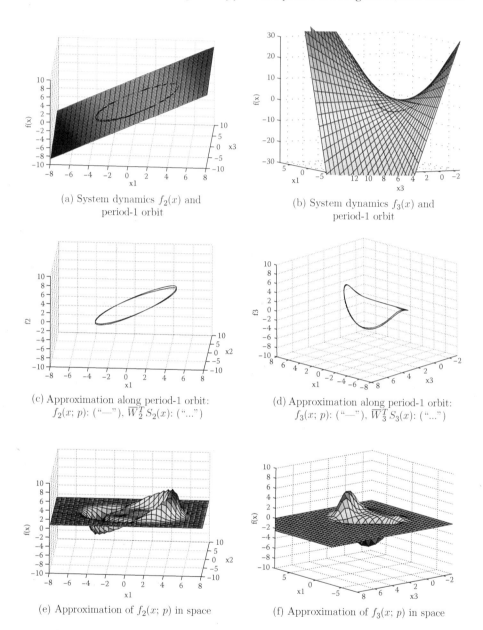

(a) System dynamics $f_2(x)$ and period-1 orbit

(b) System dynamics $f_3(x)$ and period-1 orbit

(c) Approximation along period-1 orbit: $f_2(x; p)$: ("—"), $\overline{W}_2^T S_2(x)$: ("...")

(d) Approximation along period-1 orbit: $f_3(x; p)$: ("—"), $\overline{W}_3^T S_3(x)$: ("...")

(e) Approximation of $f_2(x; p)$ in space

(f) Approximation of $f_3(x; p)$ in space

FIGURE 3.2

Approximation of system dynamics underlying a period-1 orbit.

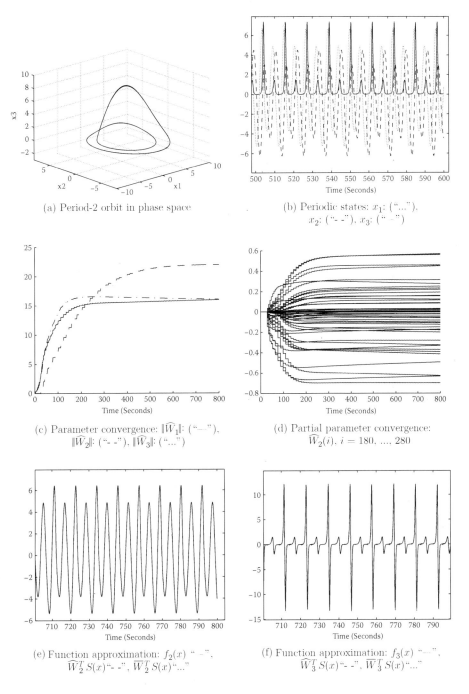

(a) Period-2 orbit in phase space

(b) Periodic states: x_1: ("..."),
x_2: ("- -"), x_3: ("- -")

(c) Parameter convergence: $\|\widehat{W}_1\|$: ("—"),
$\|\widehat{W}_2\|$: ("- -"), $\|\widehat{W}_3\|$: ("...")

(d) Partial parameter convergence:
$\widehat{W}_2(i)$, $i = 180, ..., 280$

(e) Function approximation: $f_2(x)$ "- -",
$\widehat{W}_2^T S(x)$ "- -", $\overline{W}_2^T S(x)$ "..."

(f) Function approximation: $f_3(x)$ "—",
$\widehat{W}_3^T S(x)$ "- -", $\overline{W}_3^T S(x)$ "..."

FIGURE 3.3
Identification of the Rossler system (3.30) with a period-2 orbit ($p_3 = 3.3$).

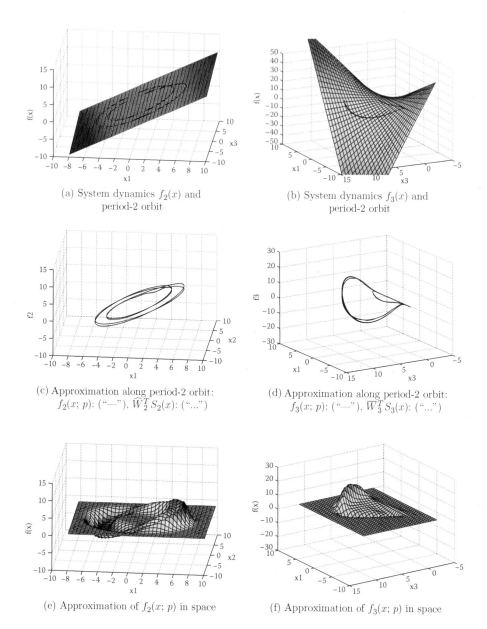

(a) System dynamics $f_2(x)$ and period-2 orbit

(b) System dynamics $f_3(x)$ and period-2 orbit

(c) Approximation along period-2 orbit: $f_2(x; p)$: ("——"), $\widehat{W}_2^T S_2(x)$: ("...")

(d) Approximation along period-2 orbit: $f_3(x; p)$: ("——"), $\overline{W}_3^T S_3(x)$: ("...")

(e) Approximation of $f_2(x; p)$ in space

(f) Approximation of $f_3(x; p)$ in space

FIGURE 3.4
Approximation of system dynamics underlying a period-2 orbit.

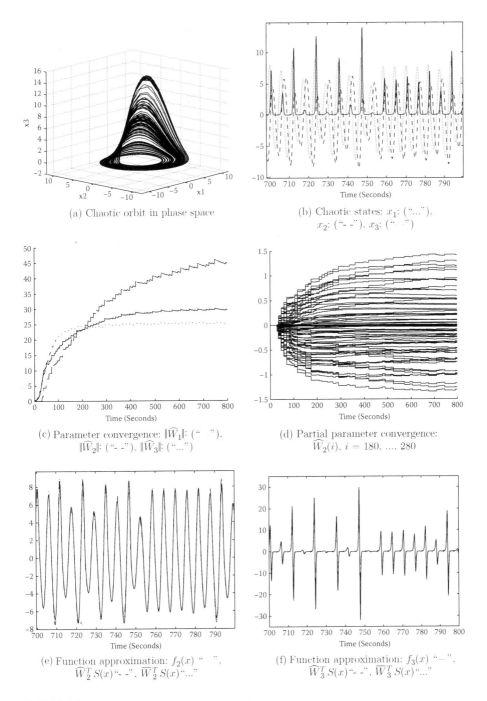

(a) Chaotic orbit in phase space

(b) Chaotic states: x_1: ("..."), x_2: ("- -"), x_3: (" ")

(c) Parameter convergence: $\|\widehat{W}_1\|$: (" "), $\|\widehat{W}_2\|$: ("- -"), $\|\widehat{W}_3\|$: ("...")

(d) Partial parameter convergence: $\widehat{W}_2(i)$, $i = 180, \ldots, 280$

(e) Function approximation: $f_2(x)$ " ", $\widehat{W}_2^T S(x)$ "- -", $\overline{W}_2^T S(x)$ "..."

(f) Function approximation: $f_3(x)$ "—", $\widehat{W}_3^T S(x)$ "- -", $\overline{W}_3^T S(x)$ "..."

FIGURE 3.5
Identification of the Rossler system (3.30) with a chaotic orbit ($p_3 = 4.5$).

space. The fast-moving trajectory leads to a much smaller τ_0, the minimum amount of time that the chaotic trajectory stays within certain neighborhoods of the involved centers. According to Remark 3.3, the smaller τ_0, plus a much larger T of the chaotic trajectory, yields much slower convergence rates of NN weight estimates $\|\widehat{W}_1\|$ and $\|\widehat{W}_3\|$. Moreover, this also leads to a much larger NN approximation error, as observed in Figures 3.5f and 3.6d.

We notice, however, that the slow parameter convergence rates do not mean that the nonlinear dynamics underlying chaotic trajectories cannot be identified. In Figure 3.7a, it is seen that within the time interval [0, 1800] seconds, all of the $\|\widehat{W}_i\|$ ($i = 1, 2, 3$) nearly converge to constant values. In Figure 3.7b, the improved NN approximation of the nonlinear dynamics $f_3(x; p)$ is shown, which yields a smaller NN approximation error compared with Figure 3.6d.

REMARK 3.5

The result clearly demonstrates that although a random-like chaotic trajectory is sensitive to initial conditions, which leads to the divergence with nearby trajectories and long-term unpredictability (called deterministic chaos), the system dynamics of a nonlinear chaotic system can still be identified along the chaotic trajectory in a deterministic way. The system dynamics underlying the chaotic trajectory is topologically similar to the dynamics underlying the two periodic trajectories. Moreover, it can be seen that identification of the system dynamics is independent of initial conditions of the periodic or chaotic trajectories. In other words, the sensitivity to initial conditions of a chaotic trajectory does not affect the identification of its underlying system dynamics. Thus, it is shown that deterministic chaos can be accurately identified via the deterministic learning approach.

REMARK 3.6

The simulations, along with the analysis in Remark 3.3, show that the slower the recurrent motions, the faster the learning speed is, and the better the accuracy of learning. On the contrary, a fast-moving trajectory may lead to a slow learning rate and a poor accuracy of learning. This is compatible with our understanding of human learning in a dynamical environment.

REMARK 3.7

Concerning the generalization issue, it is seen that NN approximation of system dynamics is valid in a local region along the recurrent trajectory. Thus, a certain ability of generalization is obtained automatically; that is, whenever the NN input comes close again to the vicinity of the experienced recurrent trajectory, the localized RBF network will provide accurate approximation to the previously learned system dynamics.

On the other hand, because NN approximation of system dynamics is invalid outside the specified trajectory, to obtain good approximations over a larger region of the space it is necessary for the NN inputs to explore a larger input space. Compared with the simple periodic trajectories, quasi-periodic

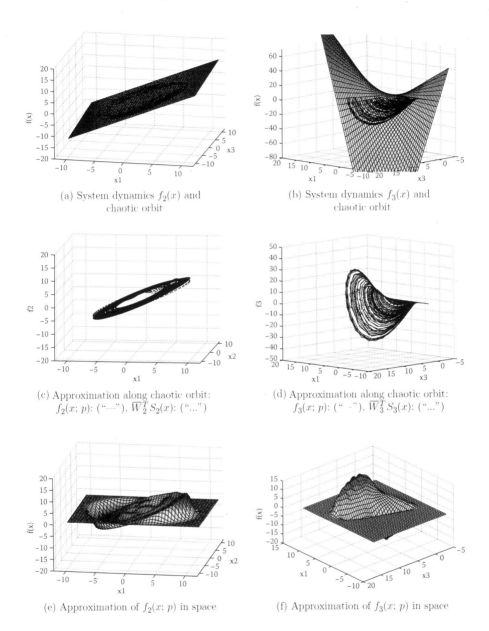

(a) System dynamics $f_2(x)$ and chaotic orbit

(b) System dynamics $f_3(x)$ and chaotic orbit

(c) Approximation along chaotic orbit: $f_2(x; p)$: ("—"), $\overline{W}_2^T S_2(x)$: ("...")

(d) Approximation along chaotic orbit: $f_3(x; p)$: (" - "), $\overline{W}_3^T S_3(x)$: ("...")

(e) Approximation of $f_2(x; p)$ in space

(f) Approximation of $f_3(x; p)$ in space

FIGURE 3.6
Approximation of system dynamics underlying a chaotic orbit.

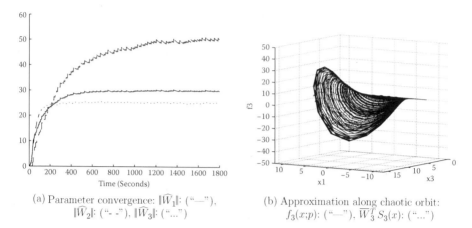

(a) Parameter convergence: $\|\widehat{W}_1\|$: ("—"),
$\|\widehat{W}_2\|$: ("- -"), $\|\widehat{W}_3\|$: ("...")

(b) Approximation along chaotic orbit:
$f_3(x;p)$: ("—"), $\overline{W}_3^T S_3(x)$: ("...")

FIGURE 3.7
Approximation of system dynamics underlying a chaotic orbit.

and chaotic trajectories are more complicated ones, because they are generally more spatially extended, which means that more neurons are involved in the regressor subvector along these trajectories. Therefore, when the nonlinear dynamical system exhibits a chaotic trajectory, the RBF networks might be better trained in the sense that better generalization ability may be obtained.

3.5   Summary

In this chapter, a "deterministic learning" mechanism has been presented, which achieves locally accurate neural network approximation of the underlying system dynamics in the local region along the recurrent trajectories. In the deterministic learning mechanism, four properties of RBF networks (linear-in-parameter, function approximation, spatially localized learning, and satisfaction of the PE condition) work together to achieve both parameter convergence and system identification in a dynamic environment. The localized RBF network is thus considered as most suitable for NN identification of nonlinear dynamical systems.

The learning is not achieved by algorithms from statistic learning theory, but is accomplished in a dynamical, deterministic manner, using results from adaptive systems theory. Specifically, with the employment of localized RBF neural networks, the recurrent trajectories of nonlinear dynamical systems lead to the satisfaction of a partial PE condition. This *a priori* verified PE condition, along with the localization property of RBF networks, yields guaranteed exponential stability of the LTV adaptive system along the recurrent trajectory. Thus, accurate learning is achieved when the corresponding NN weight estimates converge exponentially to small neighborhoods of their optimal values. The knowledge learned from deterministic learning is represented as an accurate NN approximation with constant neural weights, which is valid only in a local region along the "experienced" trajectory.

The nature of this deterministic learning is related to the exponential stability of the linear time-varying (LTV) adaptive system. It has been shown that the recurrent trajectories, which represent a large class of dynamical behaviors (or dynamical patterns) generated from nonlinear dynamical systems, including even the "unstable" chaotic ones, can all be learned and understood by deterministic learning. In other words, even for a random-like chaotic orbit that is extremely sensitive to initial conditions and is long-term unpredictable, the system dynamics can still be identified along the chaotic trajectory in a deterministic way. The proposed "deterministic learning" methodology provides an effective way for identification or modeling of nonlinear dynamical systems.

4

Deterministic Learning
from Closed-Loop Control

4.1   Introduction

In Chapter 1, we have discussed the learning issues in different areas of feedback control, including adaptive control, learning control, intelligent control, and NN control. In the discussions, the key point is that true learning ability is not typically implemented in closed-loop control systems especially in a dynamic sense. For example, although much progress has been achieved in the area of adaptive NN control (ANC), which mainly emphasizes stability and convergence of closed-loop control systems, true learning capability is actually very limited, because it still needs to recalculate (or readapt) the parameters (neural weights) even for repeating exactly the same control task. Most of the work in the ANC literature utilizes only the universal function approximation capability of neural networks to parameterize the unknown system dynamics, and is developed along the lines of well-established robust adaptive control theory [92]. ANC has only been shown to have the ability to adapt to the unknown system dynamics through online adjustment of the neural weights but does not have the ability to learn true models of system dynamics in stable closed-loop control processes.

The capability of learning knowledge online through a stable closed-loop control process requires not only the ability of adaptation to cope with system uncertainties, but also ability beyond adaptation, e.g., knowledge acquisition in dynamic environments. This kind of learning ability is related to the problem of closed-loop identification of unknown system dynamics, which has been a challenging problem in the areas of system identification and adaptive control [48]. To achieve accurate parameter convergence in closed-loop adaptive control, the persistent excitation (PE) condition of some internal closed-loop signals (rather than that of the external reference signals) is normally required to be satisfied. This is not an easy task in the control literature. Although interesting results on ANC were obtained in [46,181,195], the satisfaction of the PE condition of internal closed-loop signals has not been established.

In the preceding chapter, a deterministic mechanism is presented for learning from dynamical processes. Particularly, for nonlinear dynamical systems undergoing recurrent motions including periodic, quasi-periodic, almost-periodic, and even chaotic ones, a "deterministic learning" approach is presented that achieves locally accurate identification of the underlying system dynamics in a local region along the trajectory. In this chapter, we investigate deterministic learning in closed-loop NN control processes. We show that an appropriately designed adaptive NN controller is capable of learning closed-loop system dynamics during tracking control to a recurrent reference orbit. By using the deterministic learning mechanism, the difficulty of satisfying PE of internal closed-loop signals is overcome in two steps. In the first step, we use ANC to achieve tracking convergence of the plant states to the periodic reference states, so that the internal plant states become recurrent signals. In the second step, thanks to the tracking convergence obtained and the associated properties of localized RBF networks, a partial PE condition of internal closed-loop signals (rather than that of the external reference signals) is satisfied. Consequently, accurate identification for a partial closed-loop system dynamics is achieved in a local region along the recurrent state trajectory, and thus a true learning ability is implemented during a closed-loop feedback control process.

In the following, we start from deterministic learning for ANC of a simple second-order nonlinear system, as shown in Section 4.2. In Section 4.3, we consider learning from direct ANC of a class of strict-feedback systems. Section 4.4 investigates the learning issues in direct ANC of a class of nonlinear systems in Brunovsky form with unknown affine terms. The results of this chapter draw substantially on the recent papers [133,240,243].

4.2   Learning from Adaptive NN Control

To demonstrate the basic idea, we consider NN tracking control of the states of a second-order nonlinear system to the periodic states of a reference model.

4.2.1   Problem Formulation

Consider a second-order nonlinear system with unity control gain:

$$\begin{cases} \dot{x}_1 = x_2 \\ \dot{x}_2 = f(x) + u \end{cases} \quad (4.1)$$

where $x = [x_1, x_2]^T \in R^2$, $u \in R$ are the state variables and system input, respectively, $f(x)$ is an unknown smooth nonlinear function, and is to be approximated by RBF network $\widehat{W}^T S(Z)$ (as given in Equation [2.8]), with NN input $Z = [x_1, x_2]^T$.

Consider a second-order reference model

$$\begin{cases} \dot{x}_{d_1} = x_{d_2} \\ \dot{x}_{d_2} = f_d(x_d) \end{cases} \tag{4.2}$$

where $x_d = [x_{d_1}, x_{d_2}]^T \in R^2$ is the system state and $f_d(\cdot)$ is a known smooth nonlinear function. We denote the system orbit starting from the initial condition $x_d(0)$ as $\varphi_d(x_d(0))$ (also as φ_d for concise presentation).

ASSUMPTION 4.1
The states of the reference model remain uniformly bounded; that is, $x_d(t) \in \Omega_d$, $\forall t \geq 0$. Moreover, the system orbit φ_d is a recurrent motion.

Our objective is to develop an adaptive NN controller using a localized RBF network such that:

1. All the signals in the closed-loop system remain uniformly bounded.
2. For a desired periodic orbit $\varphi_d(x_d(0))$, generated from reference model (4.2), the state tracking error $\tilde{x} = x - x_d$ converges exponentially to an arbitrarily small neighborhood of zero in a finite time T, so that the tracking orbit $\varphi_\zeta(x(T))$ [denoted as the orbit of system (4.1) starting from $x(T)$, also as φ_ζ for conciseness] follows closely to $\varphi_d(x_d(T))$.
3. After the tracking convergence is obtained, a locally accurate approximation of $f(x)$ is achieved along the tracking orbit $\varphi_\zeta(x(T))$ by localized RBF network $\widehat{W}^T S(Z)$, as well as by $\overline{W}^T S(Z)$, where $Z = x = [x_1, x_2]^T$, $\widehat{W}$ is the vector of neural weights updated by the adaptation law given below, and $\overline{W}$ is a constant vector obtained from $\widehat{W}(t)|_{t>T}$ (given later).

REMARK 4.1
Simple as plant (4.1) is, it is noted that there appears to be no result in the NN control literature to achieve learning objective 3 above. For adaptive NN control of system (4.1), interesting results have been obtained (e.g., in [45,46]), which indicate that with locally supported basis function approximators, only PE of a reduced dimension regressor subvector will lead to the exponential stability of the closed-loop system. However, the PE condition of closed-loop signals is not shown to be satisfied, and there is not a rigorous analysis showing that accurate approximation of system dynamics can be achieved.

4.2.2 Learning from Closed-Loop Control

We present an adaptive NN controller (similar to [65]) using a Gaussian RBF network as

$$u = -z_1 - c_2 z_2 - \widehat{W}^T S(Z) + \dot{\alpha}_1 \tag{4.3}$$

where

$$z_1 = x_1 - x_{d_1} \qquad (4.4)$$

$$z_2 = x_2 - \alpha_1 \qquad (4.5)$$

$$\alpha_1 = -c_1 z_1 + \dot{x}_{d_1} = -c_1 z_1 + x_{d_2} \qquad (4.6)$$

$$\dot{\alpha}_1 = -c_1 \dot{z}_1 + \dot{x}_{d_2} = -c_1(-c_1 z_1 + z_2) + f_d(x_d) \qquad (4.7)$$

and $c_1, c_2 > 0$ are control gains. The Gaussian RBF network $\widehat{W}^T S(Z)$, defined in Equations (2.8) and (2.4), is used to approximate the unknown function $f(x)$, where $Z = x = [x_1, x_2]^T$ is the NN input, $\widehat{W}$ is the estimate of W^*, and is updated by

$$\dot{\widehat{W}} = \dot{\widetilde{W}} = \Gamma(S(Z)z_2 - \sigma \widehat{W}) \qquad (4.8)$$

where $\widetilde{W} = \widehat{W} - W^*$, $\Gamma = \Gamma^T > 0$ is a design matrix, and $\sigma > 0$ is of small value.

The following theorem indicates how both control and learning can be implemented simultaneously in the stable control process [243].

THEOREM 4.1

Consider the closed-loop system consisting of the plant (4.1), the reference model (4.2), the controller (4.3), and the NN weight updating law (4.8). For almost any recurrent orbit $\varphi_d(x_d(0))$, starting from initial condition $x_d(0) \in \Omega_d$, and with initial conditions $x(0) \in \Omega_0$ (where Ω_0 is a compact set) and $\widehat{W}(0) = 0$, we have: (i) all signals in the closed-loop system remain uniformly bounded; (ii) the state tracking error $\tilde{x}(t) = x(t) - x_d(t)$ converges exponentially to a small neighborhood around zero by appropriately choosing design parameters, and a partial persistent excitation (PE) condition of internal closed-loop signals is satisfied; and (iii) along the tracking orbit $\varphi_\zeta(x(T))$, the neural-weight estimates $\widehat{W}_\zeta$ converge to small neighborhoods of their optimal values W_ζ^*, and an accurate approximation for the unknown $f(x)$ is obtained by $\widehat{W}^T S(Z)$ to the error level ϵ (defined in Section 2.1.2), as well as by $\overline{W}^T S(Z)$, where

$$\overline{W} = mean_{t \in [t_a, t_b]} \widehat{W}(t) \qquad (4.9)$$

with $[t_a, t_b]$, $t_b > t_a > T$ representing a time segment after the transient process.

PROOF (i) Boundedness of all the signals in the closed loop can be proved similarly to results in [46,65,195]. The proof is included here for completeness.
The derivatives of z_1 and z_2 are given as below:

$$\dot{z}_1 = \dot{x}_1 - \dot{x}_{d_1} = x_2 - x_{d_2} = -c_1 z_1 + z_2 \qquad (4.10)$$

$$\dot{z}_2 = f(x) + u - \dot{\alpha}_1 = -z_1 - c_2 z_2 - \widetilde{W}^T S(Z) + \epsilon \qquad (4.11)$$

Combined with Equation (4.8), the overall closed-loop system is described by

$$
\begin{cases}
\dot{z}_1 = -c_1 z_1 + z_2 \\
\dot{z}_2 = -z_1 - c_2 z_2 - \tilde{W}^T S(Z) + \epsilon \\
\hat{W} = \tilde{W} = \Gamma(S(Z) z_2 - \sigma \hat{W})
\end{cases}
\tag{4.12}
$$

Consider the following Lyapunov function candidate:

$$
V = \frac{1}{2} z_1^2 + \frac{1}{2} z_2^2 + \frac{1}{2} \tilde{W}^T \Gamma^{-1} \tilde{W}
\tag{4.13}
$$

The derivative of V is

$$
\begin{aligned}
\dot{V} &= z_1 \dot{z}_1 + z_2 \dot{z}_2 + \tilde{W}^T \Gamma^{-1} \dot{\tilde{W}} \\
&= -c_1 z_1^2 - c_2 z_2^2 + z_2 \epsilon - \sigma \tilde{W}^T \hat{W}
\end{aligned}
\tag{4.14}
$$

Let $c_2 = c_{21} + c_{22}$ with $c_{21} = c_1 > 0$, $c_{22} > 0$. Since

$$
-c_{22} z_2^2 + z_2 \epsilon \leq \frac{\epsilon^2}{4 c_{22}} \leq \frac{\epsilon^{*2}}{4 c_{22}}
$$

$$
-\sigma \tilde{W}^T \hat{W} \leq -\sigma \|\tilde{W}\|^2 + \sigma \|\tilde{W}\| \|W^*\| \leq -\frac{\sigma \|\tilde{W}\|^2}{2} + \frac{\sigma \|W^*\|^2}{2}
$$

then Equation (4.14) becomes

$$
\dot{V} \leq -c_1 z_1^2 - c_{21} z_2^2 - \frac{\sigma \|\tilde{W}\|^2}{2} + \frac{\sigma \|W^*\|^2}{2} + \frac{\epsilon^{*2}}{4 c_{22}}
\tag{4.15}
$$

From the above, it is clear that $\dot{V}$ is negative definite whenever

$$
|z_1| > \frac{\epsilon^*}{2\sqrt{c_1 c_{22}}} + \sqrt{\frac{\sigma}{2 c_1}} \|W^*\|, \quad |z_2| > \frac{\epsilon^*}{2\sqrt{c_{21} c_{22}}} + \sqrt{\frac{\sigma}{2 c_{21}}} \|W^*\|,
$$

$$
\text{or} \quad \|\tilde{W}\| > \frac{\epsilon^*}{\sqrt{2\sigma c_{22}}} + \|W^*\|.
$$

This leads to UUB of both $z = [z_1, z_2]^T$ and $\tilde{W}$ according to

$$
|z_1| \leq \frac{\epsilon^*}{2\sqrt{c_1 c_{22}}} + \sqrt{\frac{\sigma}{2 c_1}} \|W^*\|
\tag{4.16}
$$

$$
|z_2| \leq \frac{\epsilon^*}{2\sqrt{c_{21} c_{22}}} + \sqrt{\frac{\sigma}{2 c_{21}}} \|W^*\|
\tag{4.17}
$$

$$
\|\tilde{W}\| \leq \frac{\epsilon^*}{\sqrt{2\sigma c_{22}}} + \|W^*\| := \tilde{W}^*
\tag{4.18}
$$

Because $z_1 = x_1 - x_{d_1}$ and x_{d_1} are bounded, we have that x_1 is bounded. From $z_2 = x_2 - \alpha_1$, and the boundedness of α_1 from Equation (4.6), we have that x_2 remains bounded. Using Equation (4.3), in which $\dot{\alpha}_1$ is bounded because every term in Equation (4.7) is bounded, and $S(Z)$ is bounded for all values of Z, we conclude that control u is also bounded. Thus, all the signals in the closed-loop system remain ultimately uniformly bounded.

(ii) In objective 2, we require that without the PE condition, x converges arbitrarily close to x_d in a finite time T. This finite-time convergence, rather than the asymptotic convergence as usually obtained in the literature, is important because it prevents the case that learning occurs only when time goes to infinity.

Consider the following Lyapunov function

$$V_z = \frac{1}{2}z_1^2 + \frac{1}{2}z_2^2 \tag{4.19}$$

The derivative of V_z is

$$\dot{V}_2 = z_1\dot{z}_1 + z_2\dot{z}_2$$
$$= -c_1 z_1^2 - c_2 z_2^2 + z_2\epsilon - z_2\tilde{W}^T S(Z)$$

Let $c_2 = \bar{c}_{21} + 2\bar{c}_{22}$ with $\bar{c}_{21}, \bar{c}_{22} > 0$, and let $c_1 = \bar{c}_{21}$. Since

$$-\bar{c}_{22}z_2^2 + z_2\epsilon \le \frac{\epsilon^2}{4\bar{c}_{22}} \le \frac{\epsilon^{*2}}{4\bar{c}_{22}}$$

$$-\bar{c}_{22}z_2^2 - z_2\tilde{W}^T S(Z) \le \frac{\|\tilde{W}\|^2 S^2(Z)}{4\bar{c}_{22}} \le \frac{\tilde{W}^{*2}s^{*2}}{4\bar{c}_{22}}$$

where s^* and $\tilde{W}^*$ are given in Equations (2.13) and (4.18), respectively. Then, Equation (4.15) becomes

$$\dot{V}_2 \le -c_1 z_1^2 - \bar{c}_{21}z_2^2 + \frac{\tilde{W}^{*2}s^{*2}}{4\bar{c}_{22}} + \frac{\epsilon^{*2}}{4\bar{c}_{22}} \tag{4.20}$$

Denote

$$\delta := \frac{\tilde{W}^{*2}s^{*2}}{4\bar{c}_{22}} + \frac{\epsilon^{*2}}{4\bar{c}_{22}} \tag{4.21}$$

It is clear that δ can be made arbitrarily small using large enough $\bar{c}_{22}$, that is, c_2. From Equation (4.20) we have the following inequality:

$$\dot{V}_z < -c_1 z_1^2 - \bar{c}_{21}z_2^2 + \delta$$
$$\le -2c_1 V_z + \delta \tag{4.22}$$

Let $\rho \overset{\triangle}{=} \delta/2c_1 > 0$; then (4.22) satisfies

$$0 \le V_z(t) < \rho + (V_z(0) - \rho)\exp(-2c_1 t) \tag{4.23}$$

From (4.23), we have

$$\sum_{k=1}^{2} \frac{1}{2} z_k^2 < \rho + (V_z(0) - \rho)\exp(-2c_1 t)$$

$$< \rho + V_z(0)\exp(-2c_1 t) \tag{4.24}$$

That is,

$$\sum_{k=1}^{2} z_k^2 < 2\rho + 2V_z(0)\exp(-2c_1 t) \tag{4.25}$$

which implies that given $\mu > \sqrt{2\rho} = \sqrt{\delta/c_1}$, there exists a finite time T, determined by c_1 and δ, such that for all $t \geq T$, both z_1 and z_2 satisfy

$$|z_i(t)| < \mu, \qquad i = 1, 2 \tag{4.26}$$

where μ is the size of a small residual set that can be made arbitrarily small by appropriate c_1 and c_2.

Since $z_1 = x_1 - x_{d_1}$, we know that x_1 will closely track x_{d_1}. From $z_2 = x_2 - \alpha_1 = x_2 + c_1 z_1 - x_{d_2}$, we get

$$x_2 - x_{d_2} = z_2 - c_1 z_1 \leq \mu + c_1 \mu \tag{4.27}$$

which is also a small value when μ is small. Therefore, both x_1 and x_2 will converge closely to x_{d_1} and x_{d_2} in finite time T. Because NN inputs $Z(t) = x(t) = [x_1, x_2]^T$ is made as periodic as $x_d(t)$ for all $t \geq T$, the persistent excitation (PE) condition of internal closed-loop signals, that is, the PE of a regression subvector $S_\zeta(Z(t))$ (for $t \geq T$), is proved to be satisfied according to Theorem 2.7 and Corollary 2.1. This ends the proof of (ii).

(iii) The periodicity of $Z(t)$ leads to PE of $S_\zeta(Z)$, but usually not the PE of the whole regression vector $S(Z)$. From the error system (4.11) and the adaptation law (4.8), the overall closed-loop system can be summarized in the following form:

$$\begin{bmatrix} \dot{z} \\ \dot{\widetilde{W}} \end{bmatrix} = \begin{bmatrix} A & -b S(Z)^T \\ \Gamma S(Z)b^T & 0 \end{bmatrix} \begin{bmatrix} z \\ \widetilde{W} \end{bmatrix} + \begin{bmatrix} b\epsilon \\ -\sigma \Gamma \widehat{W} \end{bmatrix} \tag{4.28}$$

where $z = [z_1, z_2]^T$, $\widetilde{W} = \widehat{W} - W^*$ are the states, A is expressed as

$$A = \begin{bmatrix} -c_1 & 1 \\ -1 & -c_2 \end{bmatrix} \tag{4.29}$$

$b = [0, 1]^T$, $\Gamma = \Gamma^T > 0$ is a constant matrix, σ is a small positive constant, ϵ is the NN approximation error bounded by ϵ^*, and $\widehat{W}$ is also bounded according to the analysis in (i).

According to the exponential convergence results as shown by Theorem 2.4, for the adaptive system (4.28), when $S(Z)$ is PE, the origin $(z, \widetilde{W}) = 0$ of the nominal system (4.28) (i.e., without the perturbation term) is exponentially stable. However, PE of $S(Z)$ requires the NN input $Z(t) = x = [x_1, x_2]^T$ to visit every center of the whole RBF network "persistently." This is not feasible in practical applications.

By using the localization property of the Gaussian RBF network, after time T, Equation (4.11) can be expressed in the following form along the tracking orbit $\varphi_\zeta(x(T))$ as:

$$\dot{z}_1 = \dot{x}_1 - \dot{x}_{d_1} = x_2 - x_{d_2} = -c_1 z_1 + z_2 \tag{4.30}$$

$$\dot{z}_2 = f(x) + u - \dot{\alpha}_1$$
$$= W_\zeta^* S_\zeta(Z) + \epsilon_\zeta - z_1 - c_2 z_2 - \widehat{W}_\zeta^T S_\zeta(Z) - \widehat{W}_{\bar{\zeta}}^T S_{\bar{\zeta}}(Z)$$
$$= -z_1 - c_2 z_2 - \widetilde{W}_\zeta^T S_\zeta(Z) + \epsilon_\zeta' \tag{4.31}$$

where $S_\zeta(Z)$ is a subvector of $S(Z)$, $\widehat{W}_\zeta$ is the corresponding weight subvector, the subscript $\bar{\zeta}$ stands for the region far away from the trajectory $\varphi_\zeta(x(T))$, with $|\widehat{W}_{\bar{\zeta}}^T S_{\bar{\zeta}}(Z)|$ being small, and $\epsilon_\zeta' = \epsilon_\zeta - \widehat{W}_{\bar{\zeta}}^T S_{\bar{\zeta}}(Z) = O(\epsilon_\zeta)$ is the NN approximation error along the trajectory φ_ζ. The closed-loop adaptive system (4.28) is now described by

$$\begin{bmatrix} \dot{z} \\ \dot{\widetilde{W}}_\zeta \end{bmatrix} = \begin{bmatrix} A & -b S_\zeta(z)^T \\ \Gamma_\zeta S_\zeta(z) b^T & 0 \end{bmatrix} \begin{bmatrix} z \\ \widetilde{W}_\zeta \end{bmatrix} + \begin{bmatrix} b \epsilon_\zeta' \\ -\sigma \Gamma_\zeta \widehat{W}_\zeta \end{bmatrix} \tag{4.32}$$

and

$$\dot{\widehat{W}}_{\bar{\zeta}} = \dot{\widetilde{W}}_{\bar{\zeta}} = \Gamma_{\bar{\zeta}}(S_{\bar{\zeta}}(z) z_2 - \sigma \widehat{W}_{\bar{\zeta}}) \tag{4.33}$$

With PE of $S_\zeta(\varphi_\zeta)$, that is, $S_\zeta(Z)$ satisfied as obtained in step (ii), according to the exponential convergence results given in Theorem 2.4, PE of $S_\zeta(Z)$ leads to the exponential stability of $(z, \widetilde{W}_\zeta) = 0$ for the nominal part of system (4.65). Since $\epsilon_\zeta' = O(\epsilon_\zeta) = O(\epsilon)$, and $\sigma \Gamma_\zeta \widehat{W}_\zeta$ can be made small by choosing σ small enough, using Theorem 2.6, both the state error $z(t)$ and the parameter error $\widetilde{W}_\zeta(t)$ converge exponentially to small neighborhoods of zero, with the sizes of the neighborhoods being determined by ϵ^* and $\sigma \|\Gamma_\zeta \widehat{W}_\zeta^*\|$.

The convergence of $\widehat{W}_\zeta$ to a small neighborhood of W_ζ^* implies that along the trajectory $\varphi_\zeta(x(T))$, we have

$$f(x) = W_\zeta^{*T} S_\zeta(Z) + \epsilon_\zeta = \widehat{W}_\zeta^T S_\zeta(Z) - \widetilde{W}_\zeta^T S_\zeta(Z) + \epsilon_\zeta$$
$$= \widehat{W}_\zeta^T S_\zeta(Z) + \epsilon_{\zeta_1} \tag{4.34}$$

where $\epsilon_{\zeta_1} = \epsilon_\zeta - \widetilde{W}_\zeta^T S_\zeta(Z) = O(\epsilon_\zeta)$ due to the convergence of $\widetilde{W}_\zeta$.

Choosing $\overline{W}$ according to Equation (4.9), Equation (4.34) can be expressed as

$$f(x) = \widehat{W}_\zeta^T S_\zeta(Z) + \epsilon_{\zeta_1}$$

$$= \overline{W}_\zeta^T S_\zeta(Z) + \epsilon_{\zeta_2} \tag{4.35}$$

where $\overline{W}_\zeta = [\overline{W}_{j_1}, \ldots, \overline{W}_{j_\zeta}]^T$ is the subvector of $\overline{W}$, and ϵ_{ζ_2} is an error using $\overline{W}_\zeta^T S_\zeta$ as the system approximation. It is clear that after the transient process, we have $\epsilon_{\zeta_2} = O(\epsilon_{\zeta_1})$.

On the other hand, for the neurons with centers far away from the trajectory φ_ζ, $|S_{\bar{\zeta}}(Z)|$ will become very small due to the localization property of Gaussian RBFs. From the adaptation law (4.33) and $\widehat{W}(0) = 0$, it can be seen that the small values of $S_{\bar{\zeta}}(\varphi_\zeta)$ will make the neural weights $\widehat{W}_{\bar{\zeta}}$ activated and updated only slightly. Both $\widehat{W}_{\bar{\zeta}}$ and $\widehat{W}_{\bar{\zeta}}^T S_{\bar{\zeta}}(x)$, as well as $\overline{W}_{\bar{\zeta}}$ and $\overline{W}_{\bar{\zeta}}^T S_{\bar{\zeta}}(x)$, will remain very small. This means that along trajectory φ_ζ, the entire RBF network $\widehat{W}^T S(Z)$ and $\overline{W}^T S(Z)$ can approximate the unknown $f(x)$ as

$$f(x) = W_\zeta^{*T} S(Z) + \epsilon_\zeta$$

$$= \widehat{W}_\zeta^T S_\zeta(Z) + \widehat{W}_{\bar{\zeta}}^T S_{\bar{\zeta}}(Z) + \epsilon_1 = \widehat{W}^T S(Z) + \epsilon_1 \qquad (4.36)$$

$$= \overline{W}_\zeta^T S_\zeta(Z) + \overline{W}_{\bar{\zeta}}^T S_{\bar{\zeta}}(Z) + \epsilon_2 = \overline{W}^T S(Z) + \epsilon_2 \qquad (4.37)$$

where $\epsilon_1 = \epsilon_{\zeta_1} - \overline{W}_{\bar{\zeta}}^T S_\zeta(z) = O(\epsilon_{\zeta_1})$, $\epsilon_2 = \epsilon_{\zeta_2} - \overline{W}_{\bar{\zeta}}^T S_\zeta(z) = O(\epsilon_{\zeta_2})$. As we also have $\epsilon_{\zeta_1} = O(\epsilon)$ and $\epsilon_{\zeta_2} = O(\epsilon)$, it is seen that both the RBF networks $\widehat{W}^T S(Z)$ and $\overline{W}^T S(Z)$ are capable of approximating the unknown nonlinearity $f(x)$ to the desired accuracy ϵ along the tracking orbit $\varphi_\zeta(x(T))$. This concludes the proof. ■

REMARK 4.2

At the end of the proof of part (i), it is clear from Equation (4.18) that no matter how we choose the design parameters, we cannot conclude any convergence result for $\|\widetilde{W}\|$. Such convergence to a small neighborhood of zero is established in part (iii). In the proof of part (iii), $\sigma \Gamma_\zeta \widehat{W}_\zeta$ can be made small since σ is chosen as a small value, and $\widehat{W}(= \widetilde{W} + W^*)$ and $\widetilde{W}$ are bounded as seen from Equation (4.18). It is important to have small ϵ_ζ' and $\sigma \Gamma_\zeta \widehat{W}_\zeta$ in Equation (4.65), so that the convergence of $\widetilde{W}_\zeta(t)$ can be guaranteed.

REMARK 4.3

In deterministic learning, the difficult problem of satisfying PE in feedback closed-loop has been overcome in two steps: (i) tracking convergence of $x(t)$ to the recurrent $x_d(t)$ in finite time T by adaptive NN control without the PE condition; and (ii) satisfaction of PE for a regression subvector $S_\zeta(Z)$ thanks to the employed RBF network, and the state tracking. In this way, the main difficulty in closed-loop identification is resolved.

REMARK 4.4

It is seen that an appropriately designed adaptive NN controller is capable of learning autonomously the system dynamics during tracking control to a recurrent reference orbit. In contrast to conventional adaptive NN control in which stability and tracking control are achieved without establishing parameter convergence, we show in this chapter that learning (i.e., parameter convergence) can be achieved from tracking control in a deterministic and autonomous way. The parameter convergence is trajectory-dependent, and the NN approximation of the closed-loop system dynamics is locally accurate along the tracking orbit. This kind of learning capability is very desirable for advanced intelligent control systems.

REMARK 4.5

From Equations (4.36) and (4.37), it is seen that although useful knowledge is obtained only in $\widehat{W}_\zeta$, it is not necessary to specify which neural weights belong to $\widehat{W}_\zeta$ and need to be updated. It is clear that the locally accurate NN approximation is achieved by using the entire RBF network $\widehat{W}^T S(Z)$ and stored in the constant RBF network $\overline{W}^T S(Z)$ for the system's uncertain non-linearity $f(x)$. This NN approximation is not valid within the entire regular lattice upon which the RBF network is constructed, but only applies in the local region along the tracking orbit $\varphi_\zeta(x(T))$. For the (possibly large) area where the tracking orbit does not explore, no learning occurs, as represented by the slightly updated $\widehat{W}_{\bar\zeta}$ (4.33). Because $S(Z)$ is of small value when $Z(t)$ is far away from the tracking orbit $\varphi_\zeta(x(T))$, together with the small values of $\widehat{W}_{\bar\zeta}$ and $\overline{W}_{\bar\zeta}$, we have $\widehat{W}^T S(Z)$ and $\overline{W}^T S(Z)$ remain small in the unexplored area. This means that nothing can be learned without sufficient "experiences."

Therefore, the learned knowledge can be interpreted as: for the experienced recurrent orbit $\varphi_d(x_d(0))$, there exist constants d, $\epsilon_2^* > 0$, which describe a local region Ω_{φ_d} along $\varphi_d(x_d(0))$, such that

$$\text{dist}(x, \varphi_d) < d \Rightarrow \left|\overline{W}^T S(x) - f(x)\right| < \epsilon_2^* \tag{4.38}$$

where ϵ_2^* is close to ϵ^*.

For a new control task, the knowledge represented in Equation (4.38) can be recalled in such a way that whenever the NN input $Z = x = [x_1, x_2]^T$ comes close again to the vicinity of the experienced tracking orbit $\varphi_d(x_d(0))$, the RBF network $\overline{W}^T S(x)$ will provide an accurate approximation to the uncertain nonlinearity.

REMARK 4.6

The system considered in this chapter is simple. It is chosen as adequate to demonstrate the proposed deterministic learning mechanism. Continued efforts are being made to investigate learning from direct adaptive NN control of more general nonlinear systems in the following sections.

4.2.3   Simulation Studies

To verify and test the proposed NN control and learning approach, the following van der Pol oscillator [28,227] is taken as the plant for control:

$$\dot{x}_1 = x_2$$
$$\dot{x}_2 = -x_1 + \beta\left(1 - x_1^2\right)x_2 + u \tag{4.39}$$

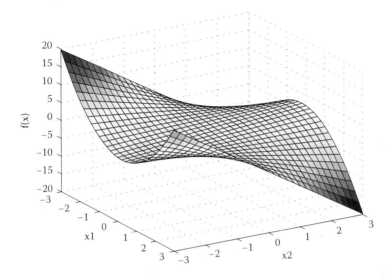

FIGURE 4.1
System nonlinearity: $f(x)$.

where $\beta > 0$ is a system parameter ($\beta = 0.7$ here); the smooth function $f(x_1, x_2) = -x_1 + \beta(1 - x_1^2)x_2$ is assumed to be unknown to the controller u. The nonlinearity of $f(x_1, x_2)$ is shown in Figure 4.1.

The desired trajectory y_d is generated from the following Duffing oscillator [28,40]:

$$\dot{x}_{d_1} = x_{d_2}$$
$$\dot{x}_{d_2} = -p_2 x_{d_1} - p_3 x_{d_1}^3 - p_1 x_{d_2} + q\cos(wt) \tag{4.40}$$

where x_{d_1} and x_{d_2} are system states; p_1, p_2, p_3 are system parameters. As shown in [28], for $p_1 = 0.4$, $p_2 = -1.1$, $p_3 = 1.0$, $w = 1.8$, the phase-plane trajectory of the Duffing oscillator approaches a period-1 limit cycle when $q = 0.620$ (as seen in Figure 4.2a). The phase-plane trajectory becomes a period-2 limit cycle when $p_1 = 0.55$ and $q = 1.498$ (as seen in Figure 4.3a). It becomes a chaotic orbit when $p_1 = 0.35$ and $q = 1.498$ (as seen in Figure 4.4a).

The Gaussian RBF network $\widehat{W}^T S(Z)$ contains 441 nodes (i.e., $N = 441$). The centers μ_i ($i = 1, \ldots, N$) are evenly spaced on $[-3.0, 3.0] \times [-3.0, 3.0]$, with widths $\eta_i = 0.3$ ($i = 1, \ldots, N$). The adaptive NN controller (4.3) is used to control the uncertain system (4.39). The weights of the NN are updated online according to Equation (4.8). The design parameters of the above controller are $c_1 = 3$, $c_2 = 10$, $\Gamma = diag\{5.0\}$, and $\sigma = 0.001$. The initial weights $\widehat{W}(0) = 0.0$, the initial conditions $[x_1(0), x_2(0)]^T = [0.5, 0.2]^T$, and $[x_{d_1}(0), x_{d_2}(0)]^T = [0.2, 0.3]^T$.

First, the period-1 signal is employed as the reference signal for training the RBF network. From Figure 4.2a, we can see that tracking of the system states to a small neighborhood of the period-1 reference orbit is achieved. The partial parameter convergence is shown in Figure 4.2b, which reveals that only part

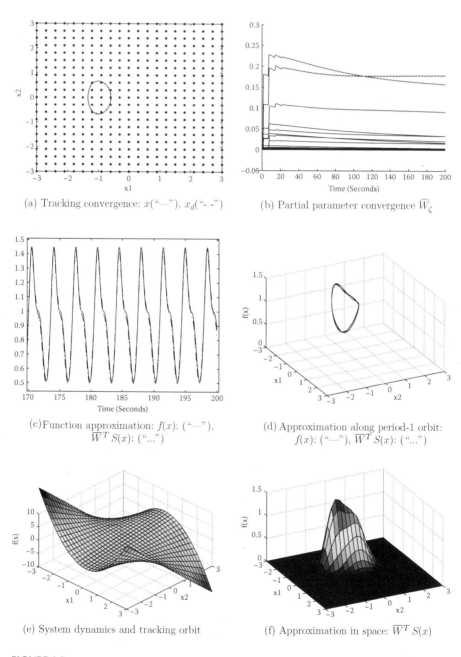

(a) Tracking convergence: x ("—"), x_d ("- -")

(b) Partial parameter convergence $\widehat{W}_\zeta$

(c) Function approximation: $f(x)$: ("—"), $\overline{W}^T S(x)$: ("...")

(d) Approximation along period-1 orbit: $f(x)$: ("—"), $\overline{W}^T S(x)$: ("...")

(e) System dynamics and tracking orbit

(f) Approximation in space: $\overline{W}^T S(x)$

FIGURE 4.2
Responses for tracking control to period-1 orbit.

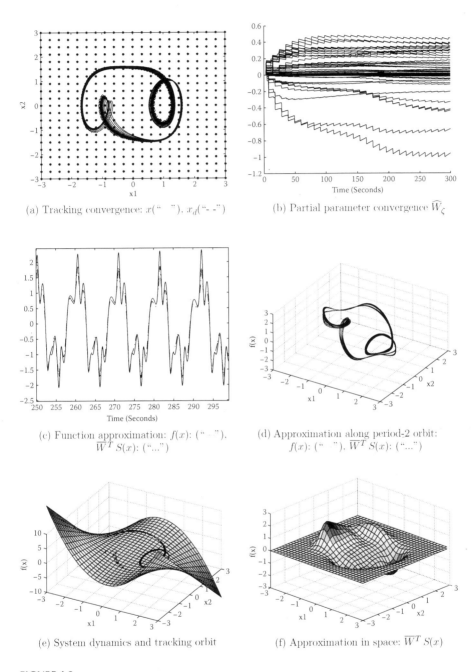

(a) Tracking convergence: x ("—"), x_d ("- -")

(b) Partial parameter convergence $\widehat{W}_\zeta$

(c) Function approximation: $f(x)$: ("—"), $\overline{W}^T S(x)$: ("...")

(d) Approximation along period-2 orbit: $f(x)$: ("—"), $\overline{W}^T S(x)$: ("...")

(e) System dynamics and tracking orbit

(f) Approximation in space: $\overline{W}^T S(x)$

FIGURE 4.3
Responses for tracking control to period-2 orbit.

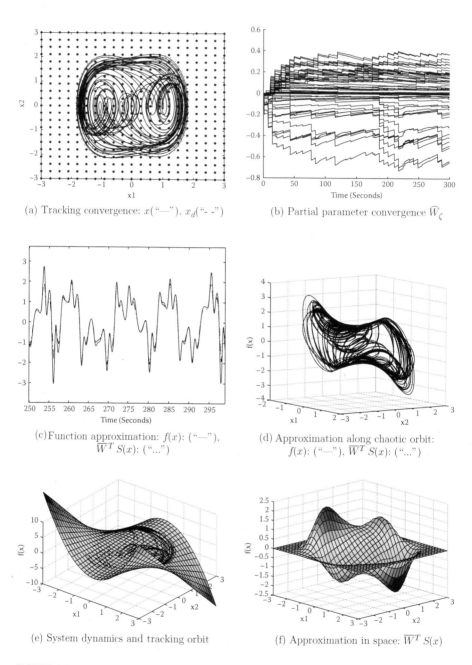

(a) Tracking convergence: x ("—"), x_d ("- -")

(b) Partial parameter convergence $\widehat{W}_\zeta$

(c) Function approximation: $f(x)$: ("—"), $\overline{W}^T S(x)$: ("...")

(d) Approximation along chaotic orbit: $f(x)$: ("—"), $\overline{W}^T S(x)$: ("...")

(e) System dynamics and tracking orbit

(f) Approximation in space: $\overline{W}^T S(x)$

FIGURE 4.4
Responses for tracking control to chaotic orbit.

of the neural weights converges; many other neural weights remain zero or small values. Because the optimal values W_ζ^* are generally unknown, it is difficult to verify whether $\hat{W}_\zeta$ have indeed converged to W_ζ^*. Fortunately, we can show the NN approximation of system dynamics $f(x)$ both in time domain and in phase space, as in Figures 4.2c and d. In Figure 4.2e, we plot the system dynamics and the tracking orbit together. Corresponding to Figures 4.2d and e, it is seen from Figure 4.2f that good NN approximation of the unknown $f(x)$ is achieved by using constant RBF network $\overline{W}^T S(x)$ along the period-1 tracking trajectory.

To obtain good approximation over a larger space, it is necessary for the NN inputs to explore a larger input space. We demonstrate such exploration using a period-2 reference orbit in Figure 4.3, and using a chaotic reference orbit in Figure 4.4. As shown in Figures 4.3a and b and Figures 4.4a and b, both tracking control and partial parameter convergence are achieved. In comparison with Figure 4.2b, it can be seen in Figures 4.3b and 4.4b that more neurons are being activated and updated. It is clearly seen from Figures 4.3d and f and Figures 4.4d and f that fairly good NN approximation of the system dynamics $f(x)$ (shown in Figures 4.3e and 4.4e) can still be obtained along the period-2 and chaotic orbits.

Figures 4.2f, 4.3f, and 4.4f clearly illustrate the knowledge representation. It is shown in Figure 4.2d that the NN approximation by $\overline{W}^T S(x)$ is only accurate in the vicinity of the period-1 orbit, rather than within the entire space of interest. For the large region where the tracking orbit does not explore, no learning occurs, corresponding to the zero-plane in Figure 4.2f, due to the small values of $\overline{W}^T S(x)$ in that area.

In the case of tracking to the period-2 and chaotic orbits, the local knowledge represented by $\overline{W}^T S(x)$ is more clearly demonstrated. As seen from Figures 4.3f and 4.4f, what is actually learned and stored in $\overline{W}^T S(x)$ is the approximation of system dynamics $f(x)$ in a local region along the period-2 and chaotic orbits. It is interesting to notice that the learned knowledge consists of "hills and valleys" outlined by the tracking orbits.

4.3 Learning from Direct Adaptive NN Control of Strict-Feedback Systems

As system (4.1) is so simple, it is necessary to extend this learning result to more general nonlinear systems. In this section, we investigate the learning issues in direct adaptive NN control of nonlinear systems in the strict-feedback form [119]. Direct ANC (e.g., [65,124,195,269]) refers to the approach in which NNs are employed to approximate the unknown dynamics in certain desired controllers, whereas in the indirect ANC approach (e.g., [46,181]), NNs are used to approximate the unknown system dynamics in the plant. Note that due to the simplicity of system (4.1), both the direct and indirect ANC approaches are applicable to achieve learning from neural control. For more general nonlinear systems, we investigate whether the deterministic learning ability can be achieved by *direct* adaptive NN control.

To implement learning from adaptive NN control, a requirement here is that all of the NN inputs become a periodic or periodic-like (recurrent) orbit such that a partial PE condition is satisfied. In direct ANC of general nonlinear systems, intermediate variables are usually introduced as NN inputs for the purpose of keeping the dimension of NN inputs minimal [65,266]. However, the introduction of intermediate variables will yield a problem concerning learning; that is, these intermediate variables are required to become periodic or periodic-like to satisfy the PE condition. This is a new requirement, and its satisfaction is the key to deterministic learning. For direct ANC of a class of general nonlinear systems in the strict-feedback form, we show that all the internal system states and the intermediate variables can still be made periodic or periodic-like along with the reference system states. Therefore, the PE condition can still be satisfied by using localized RBF networks, and accurate learning of control system dynamics can be achieved from a direct ANC process.

4.3.1   Problem Formulation

Consider the following nonlinear system in the strict-feedback form [119]

$$\begin{cases} \dot{x}_1 = f_1(x_1) + x_2 \\ \dot{x}_2 = f_2(x_1, x_2) + u \end{cases} \tag{4.41}$$

where $x = [x_1, x_2]^T \in R^2$, $u \in R$ are the state variables and system input, respectively, and $f_1(x_1)$ and $f_2(x_1, x_2)$ are both unknown but smooth nonlinear functions.

Consider the following smooth, bounded reference model

$$\begin{aligned} \dot{x}_{di} &= f_{di}(x_d), \qquad 1 \le i \le 2 \\ y_d &= x_{d_1} \end{aligned} \tag{4.42}$$

where $x_d = [x_{d_1}, x_{d_2}]^T \in R^2$ are the states, $y_d \in R$ is the system output, and $f_{di}(\cdot)$, $i = 1, 2$ are unknown smooth nonlinear functions. Assume that both $x_{d_1} (= y_d)$ and x_{d_2} are periodic signals or periodic-like recurrent and the reference orbit [denoted as $\varphi_d(x_d(0))$ or φ_d] is a periodic motion.

The objective is to develop a direct adaptive NN controller using localized RBF networks such that:

1. All the signals in the closed-loop system remain uniformly bounded.

2. The output y of system (4.41) converges exponentially to a desired trajectory y_d generated from Equation (4.42), such that the output tracking error $y - y_d$ converges to a small neighborhood of zero in a finite time T.

3. The unknown control system dynamics are accurately approximated by localized RBF networks along trajectories of NN inputs.

REMARK 4.7

For adaptive NN control (ANC) of system (4.41), the direct ANC approach (e.g., [65]) employs NNs to approximate the unknown nonlinearity $h(x, v)$ in the desired control u^*, where $h(\cdot)$ is the unknown control system dynamics; v is a vector of some intermediate variables. The indirect ANC approach, on the other hand, uses NNs to identify the system nonlinearities $f_1(x_1)$ and $f_2(x_1, x_2)$ (e.g., see [181]). For ANC of general nonlinear systems, it is normally considered that the direct approach provides a better solution than the indirect approach [269]. However, the learning issue in both approaches, that is, accurate learning of either $h(x, v)$ or $f_i(\cdot)$ ($i = 1, 2$), has not previously been fully studied.

4.3.2 Direct ANC Design

For the control of strict-feedback system (4.41), the direct ANC approach developed in [65] is applicable. At each recursive step i ($i = 1, 2$), a desired feedback control α_i^* is first shown to exist. Then, a stabilizing function α_i ($u = \alpha_2$) is designed, where a localized RBF network is employed to approximate the unknown nonlinearity in α_i^* ($i = 1, 2$).

STEP 4.1

Define $z_1 = x_1 - x_{d_1}$. Its derivative is $\dot{z}_1 = f_1(x_1) + x_2 - \dot{x}_{d_1}$. By viewing x_2 as a virtual control input, it is clear that there exists a desired virtual control $\alpha_1^* \triangleq x_2$,

$$\alpha_1^* = -c_1 z_1 - f_1(x_1) + \dot{x}_{d_1}$$

where $c_1 > 0$ is a design constant.

Denote $h_1(Z_1) \triangleq f_1(x_1)$, where $Z_1 \triangleq [x_1]^T \in \Omega_1 \subset R$. By employing an RBF neural network $W_1^T S_1(Z_1)$ to approximate $h_1(Z_1)$ in a compact set Ω_1, we have

$$h_1(Z_1) = W_1^{*T} S_1(Z_1) + \epsilon_1, \qquad \forall Z_1 \in \Omega_1 \tag{4.43}$$

where W_1^* denotes the ideal constant weights, and $|\epsilon_1| \le \epsilon_1^*$ is the approximation error with constant $\epsilon_1^* > 0$. Let $\widehat{W}_1$ be the estimate of W_1^* and $\widetilde{W}_1 = \widehat{W}_1 - W_1^*$.

Define $z_2 = x_2 - \alpha_1$ and let

$$\alpha_1 = -c_1 z_1 - \widehat{W}_1^T S_1(Z_1) + \dot{x}_{d_1} \tag{4.44}$$

where $\widehat{W}_1$ is updated by

$$\dot{\widehat{W}}_1 = \dot{\widetilde{W}}_1 = \Gamma_1 S_1(Z_1) z_1 - \sigma_1 \Gamma_1 \widehat{W}_1 \tag{4.45}$$

with $\Gamma_1 = \Gamma_1^T > 0$ and $\sigma_1 > 0$ being a small constant.

Then, the dynamics of z_1 are governed by

$$\dot{z}_1 = f_1(x_1) + (z_2 + \alpha_1) - \dot{x}_{d_1}$$
$$= -c_1 z_1 + z_2 - \widetilde{W}_1^T S_1(Z_1) + \epsilon_1 \tag{4.46}$$

STEP 4.2

The derivative of $z_2 = x_2 - \alpha_1$ is $\dot{z}_2 = f_2(x_1, x_2) + u - \dot{\alpha}_1$.

To stabilize the (z_1, z_2)-system, there exists a desired feedback control

$$u^* = -z_1 - c_2 z_2 - (f_2(x_1, x_2) - \dot{\alpha}_1) \tag{4.47}$$

where $c_2 > 0$ is a design constant. From Equation (4.44), it can be seen that α_1 is a function of x_1, x_d, and $\widehat{W}_1$. Thus, $\dot{\alpha}_1$ is given by

$$\dot{\alpha}_1 = \frac{\partial \alpha_1}{\partial x_1} \dot{x}_1 + \frac{\partial \alpha_1}{\partial x_d} \dot{x}_d + \frac{\partial \alpha_1}{\partial \widehat{W}_1} \dot{\widehat{W}}_1$$

$$= \frac{\partial \alpha_1}{\partial x_1} (f_1(x_1) + x_2) + \phi_1 \tag{4.48}$$

where

$$\phi_1 = \frac{\partial \alpha_1}{\partial x_d} \dot{x}_d + \frac{\partial \alpha_1}{\partial \widehat{W}_1} [\Gamma_1(S_1(Z_1)z_1 - \sigma_1 \Gamma_1 \widehat{W}_1)]$$

is computable.

Let

$$h_2(Z_2) \triangleq \left[f_2(x_1, x_2) - \frac{\partial \alpha_1}{\partial x_1} (f_1(x_1) + x_2) \right] \tag{4.49}$$

where

$$Z_2 \triangleq \left[x_1, x_2, \frac{\partial \alpha_1}{\partial x_1} \right]^T \in \Omega_2 \subset R^3 \tag{4.50}$$

By employing an RBF network $W_2^T S_2(Z_2)$ to approximate $h_2(Z_2)$ within Ω_2, we have

$$h_2(Z_2) = W_2^{*T} S_2(Z_2) + \epsilon_2, \qquad \forall Z_2 \in \Omega_2 \tag{4.51}$$

where W_2^* denotes the ideal constant weights, and $|\epsilon_2| \le \epsilon_2^*$ is the approximation error with constant $\epsilon_2^* > 0$.

Choose the practical control

$$u = -z_1 - c_2 z_2 - \widehat{W}_2^T S_2(Z_2) + \phi_1 \tag{4.52}$$

where $\widehat{W}_2$ is updated by

$$\dot{\widehat{W}}_2 = \dot{\widetilde{W}}_2 = \Gamma_2 S_2(Z_2)z_2 - \sigma_2 \Gamma_2 \widehat{W}_2 \tag{4.53}$$

with $\Gamma_2 = \Gamma_2^T > 0$ and $\sigma_2 > 0$ being a small constant.

Then, we have

$$\dot{z}_2 = f_2(x_1, x_2) + u - \dot{\alpha}_1$$

$$= -z_1 - c_2 z_2 - \widetilde{W}_2^T S_2(Z_2) + \epsilon_2 \tag{4.54}$$

REMARK 4.8

By defining intermediate variable $\frac{\partial \alpha_1}{\partial x_1}$, which is available through the computation of x_1, x_d and $\widehat{W}_1$, the NN approximation $\widehat{W}_2^T S_2(Z_2)$ of the unknown function $h_2(Z_2)$ can be computed by using the minimal number of NN inputs $Z_2 = [x_1, x_2, \frac{\partial \alpha_1}{\partial x_1}]^T$.

THEOREM 4.2 **(Stability and Tracking)**

Consider the closed-loop system consisting of the plant (4.41), the reference model (4.42), the controller (4.52), and the NN weight updating laws (4.45) and (4.53). For sufficiently large compact sets Ω_1 and Ω_2, with initial conditions appropriately chosen, and with $\widehat{W}(0) = 0$, we have that: (i) all the signals in the closed-loop system remain bounded, and (ii) the output tracking error $y(t) - y_d(t)$ converges to a small neighborhood around zero for all $t \geq T$ by appropriately choosing design parameters.

PROOF The system (4.41) is a simple case of the class of strict-feedback systems considered in [65]. Thus, the stability of all the signals in the closed-loop system, including z_1, z_2, x_1, x_2, $\widehat{W}_1$, $\widehat{W}_2$, α_1, $\dot{\alpha}_1$, and u, can be easily concluded as in [65].

Similar to the proof of Theorem 4.1, it can be derived that by choosing large c_1 and c_2, both z_1 and z_2 will converge exponentially to a small neighborhood of zero. Therefore, there exists a time $T > 0$, such that for all $t \geq T$, the output tracking error $y(t) - y_d(t)$ converges to a small neighborhood of zero. ■

4.3.3 Learning from Direct ANC

To achieve the learning objective (iii), that is, accurate NN approximation of $h_i(Z)$ using $\widehat{W}_i^T S_i(Z_i)$ along the trajectories of NN inputs $Z_i(t)$, it is required that the PE condition of regression subvectors along the trajectory $Z_i(t)$, that is, PE of $S_{1\zeta}(Z_1)$ and $S_{2\zeta}(Z_2)$, be satisfied.

In Section 4.2, PE of $S_\zeta(Z)$ is satisfied thanks to (a) the associated properties of the localized RBF networks and (b) the obtained tracking convergence which makes the internal system states $x(t)$ (and the NN inputs $Z = [x_1, x_2]^T$) follow a desired recurrent trajectory $x_d(t)$. For direct ANC of system (4.41), apart from the tracking convergence of x_1 to x_{d_1}, it is required to make both x_2 and $\frac{\partial \alpha_1}{\partial x_1}$ recurrent, as is the system state x_1.

In the following, we will show that accurate learning of control system dynamics $h_i(Z_i)$ can still be achieved, and it is indeed possible to implement learning from direct ANC of strict-feedback systems.

THEOREM 4.3 **(Learning)**

Consider the closed-loop system consisting of the plant (4.41), the reference model (4.42), the controller (4.52), and the NN weight updating laws (4.45) and (4.53). For almost any recurrent orbit $\varphi_d(x_d(0))$, and with initial conditions $x(0) \in \Omega_0$

(where Ω_0 is an appropriately chosen compact set) and $\widehat{W}_i(0) = 0$, we have that (i) along the NN input orbit $Z_1(t)$ $(t > T)$, neural-weight estimates $\widehat{W}_{1\zeta}$ converge to small neighborhoods of their optimal values $W_{1\zeta}^*$, and accurate approximation for the control dynamics $h_1(Z_1)$ is obtained by $\widehat{W}_1^T S_1(Z_1)$ and $\overline{W}_1^T S_1(Z_1)$, where $\overline{W}_1$ is obtained from $\widehat{W}_1$ according to Equation (4.9). (ii) Along the NN input orbit $Z_2(t)$ $(t > T_1 > T)$, neural-weight estimates $\widehat{W}_{2\zeta}$ converge to small neighborhoods of their optimal values $W_{2\zeta}^*$, and accurate approximation for the control dynamics $h_2(Z_2)$ is obtained by $\widehat{W}_2^T S_2(Z_2)$ and $\overline{W}_2^T S_2(Z_2)$, where $\overline{W}_2$ is a constant vector obtained from $\widehat{W}_2$.

PROOF   (i) With the boundedness of all the signals in the closed-loop system, and with the exponential convergence of both $z_1 = x_1 - x_{d_1}$ and $z_2 = x_2 - \alpha_1$, (as established in Theorem 4.2), we have that x_1 converges closely to the recurrent x_{d_1} for all $t > T$. Therefore, the NN input $Z_1 = [x_1]^T$ will follow a recurrent orbit for all $t \geq T$, and consequently, a partial PE condition of $S_{1\zeta}(Z_1)$ will be satisfied.

By using the localized RBF network, along the tracking orbit $Z_1(t)$ $(t > T)$, the closed-loop adaptive subsystem, including Equations (4.45), and (4.46), can be expressed as:

$$\dot{z}_1 = -c_1 z_1 - \widetilde{W}_{1\zeta}^T S_{1\zeta}(Z_1) + z_2 + \epsilon_{1\zeta}' \tag{4.55}$$

$$\dot{\widehat{W}}_{1\zeta} = \dot{\widetilde{W}}_{1\zeta} = \Gamma_{1\zeta} S_{1\zeta}(Z_1) z_1 - \sigma_1 \Gamma_\zeta \widehat{W}_{1\zeta} \tag{4.56}$$

and

$$\dot{\widehat{W}}_{\bar{1}\zeta} = \dot{\widetilde{W}}_{\bar{1}\zeta} = \Gamma_{\bar{1}\zeta} S_{\bar{1}\zeta}(Z_1) z_1 - \sigma_1 \Gamma_{\bar{1}\zeta} \widehat{W}_{\bar{1}\zeta} \tag{4.57}$$

where $S_{1\zeta}(Z_1)$ is a subvector of $S_1(Z_1)$ as defined in Equation (2.12), $\widehat{W}_{1\zeta}$ is the corresponding weight subvector, the subscript $(\cdot)_{\bar{1}\zeta}$ stands for the region far away from the trajectory $Z_1(t)$, with $|\widehat{W}_{\bar{1}\zeta}^T S_{\bar{1}\zeta}(Z)|$ being small, and $\epsilon_{1\zeta}' = \epsilon_1 - \widetilde{W}_{\bar{1}\zeta}^T S_{\bar{1}\zeta}(Z_1) = O(\epsilon_1)$ is the NN approximation error along the trajectory $Z_1(t)$.

With PE of $S_{1\zeta}(Z_1)$, it is concluded according to Theorem 2.4 that exponential stability of $(z_1, \widetilde{W}_{1\zeta}) = 0$ for the nominal part of system (4.55) and (4.56) can be achieved. Then, $z_1(t)$, and especially $\widetilde{W}_{1\zeta}(t)$ will converge exponentially to small neighborhoods of zero, with the sizes of the neighborhoods being determined by ϵ_1^*, z_2, and $\sigma \Gamma_{1\zeta} \widehat{W}_{1\zeta}^*$, where $|z_2|$ has been shown to converge to a small neighborhood of zero.

The convergence of $\widehat{W}_{1\zeta}$, together with the localization property of RBFs, implies that along $Z_1(t)$ $(t > T)$, the control system dynamics $h_1(Z_1)$ can be accurately approximated by $\widehat{W}_{1\zeta}^T S_{1\zeta}(Z_1)$ and the entire RBF network $\widehat{W}_1^T S_1(Z_1)$ as

$$h_1(Z_1) = \widehat{W}_{1\zeta}^T S_{1\zeta}(Z_1) + \epsilon_{\zeta_{11}} \tag{4.58}$$

$$= \widehat{W}_1^T S_1(Z_1) + \epsilon_{11} \tag{4.59}$$

where $\epsilon_{\zeta_{11}} = O(\epsilon_1)$ and $\epsilon_{11} = O(\epsilon_1)$ due to the convergence of $\widetilde{W}_{1\zeta}$.

Choosing $\overline{W}_1$ according to Equation (4.9), along the trajectory $Z_1(t)$ accurate approximation for the unknown $h_1(Z_1)$ is also obtained by using $\overline{W}_{1\zeta}^T S_{1\zeta}(Z)$ and $\overline{W}_1^T S_1(Z_1)$; that is,

$$h_1(Z_1) = \overline{W}_{1\zeta}^T S_{1\zeta}(Z) + \bar{\epsilon}_{\zeta_{12}} \tag{4.60}$$

$$= \overline{W}_1^T S_1(Z_1) + \bar{\epsilon}_{12} \tag{4.61}$$

where $\epsilon_{\zeta_{12}} = O(\epsilon_1)$ and $\epsilon_{12} = O(\epsilon_1)$, respectively, after the transient process.

(ii) To achieve learning of $h_2(Z_2)$, we require both x_2 and $\frac{\partial \alpha_1}{\partial x_1}$ to become periodic or periodic-like signals. Since $x_2 = z_2 + \alpha_1$, and $\alpha_1 = -c_1(x_1 - x_{d_1}) + \widehat{W}_1^T S_1(Z_1)$, with the exponential convergence of $\widehat{W}_{1\zeta}$ to $\overline{W}_{1\zeta}$, there exists a constant $T_1 > T$ such that

$$x_2 = -c_1(x_1 - x_{d_1}) + \overline{W}_1^T S_1(Z_1) + z_2 + \varepsilon_{11} \tag{4.62}$$

holds for all $t > T_1$, where $\varepsilon_{11} = \widehat{W}_1^T S_1(Z_1) - \overline{W}_1^T S_1(Z_1)$, and both $|z_2|$ and $|\varepsilon_{11}|$ are small values. Thus, x_2 becomes a periodic-like signal, with the same period as x_1 and x_{d1}.

Furthermore, the intermediate variable

$$\frac{\partial \alpha_1}{\partial x_1} = -c_1 + \widehat{W}_1^T \frac{\partial S_1(Z_1)}{\partial x_1}$$

$$= -c_1 + \overline{W}_1^T \frac{\partial S_1(Z_1)}{\partial x_1} + \varepsilon_{12}, \qquad \forall t > T_1 \tag{4.63}$$

where $\varepsilon_{12} = \widehat{W}_1^T \frac{\partial S_1(Z_1)}{\partial x_1} - \overline{W}_1^T \frac{\partial S_1(Z_1)}{\partial x_1}$ is small, will become a periodic-like signal with the same period as x_1 for all $t > T_1$. Therefore, the NN inputs $Z_2 = [x_1, x_2, \frac{\partial \alpha_1}{\partial x_1}]^T$ will follow a periodic-like orbit for all $t > T_1$, and consequently, from Corollary 2.1, a partial PE condition of $S_{2\zeta}(Z_2)$ will be satisfied.

By using the localization property of RBF networks, along the tracking orbit $Z_2(t)$ $(t > T_1 > T)$, the closed-loop adaptive subsystem, including (4.53) and (4.54), can be expressed as:

$$\dot{z}_2 = -c_2 z_2 - \widetilde{W}_{2\zeta}^T S_{2\zeta}(Z_2) - z_1 + \epsilon'_{2\zeta} \tag{4.64}$$

$$\dot{\widehat{W}}_{2\zeta} = \dot{\widetilde{W}}_{2\zeta} = \Gamma_{2\zeta} S_{2\zeta}(Z_2) z_2 - \sigma_2 \Gamma_{2\zeta} \widehat{W}_{2\zeta} \tag{4.65}$$

and

$$\dot{\widehat{W}}_{2\bar{\zeta}} = \dot{\widetilde{W}}_{2\bar{\zeta}} = \Gamma_{2\bar{\zeta}} S_{2\bar{\zeta}}(Z_2) z_2 - \sigma_2 \Gamma_{2\bar{\zeta}} \widehat{W}_{2\bar{\zeta}} \tag{4.66}$$

where $\epsilon'_{2\zeta} = \epsilon_2 - \widetilde{W}_{2\bar{\zeta}}^T S_{2\bar{\zeta}}(Z_2) = O(\epsilon_2)$ is the NN approximation error along the trajectory $Z_2(t)$.

With PE of $S_{2\zeta}(Z_2)$, it is concluded that exponential stability of $(z_2, \widetilde{W}_{2\zeta}) = 0$ for the nominal part of system (4.64) and (4.65) can be achieved [161]. Then,

$\widetilde{W}_{2\zeta}(t)$ will converge exponentially to small neighborhoods of zero, with the sizes of the neighborhoods being determined by ϵ_2^*, $|z_1|$, and $\sigma_2\Gamma_{2\zeta}\widehat{W}_{2\zeta}^*$, where z_1 has been shown to converge to a small neighborhood of zero.

Similarly to step (i), it can be concluded that along $Z_2(t)$ $(t > T_1)$, the control system dynamics $h_2(Z_2)$ can be accurately approximated by $\widehat{W}_2^T S_2(Z_2)$ and $\overline{W}_2^T S_2(Z_2)$ as

$$h_2(Z_2) = \widehat{W}_2^T S_2(Z_2) + \epsilon_{21} \tag{4.67}$$

$$= \overline{W}_2^T S_2(Z_2) + \epsilon_{22} \tag{4.68}$$

where $\overline{W}_2$ is chosen according to (4.9), and $\epsilon_{21} = O(\epsilon_2)$, $\epsilon_{22} = O(\epsilon_2)$. This ends the proof. ∎

REMARK 4.9

Following the principle of making the NN inputs become a periodic or periodic-like orbit in the NN input space, we achieve deterministic learning from direct ANC of a more general nonlinear system (4.41) than treated in Section 4.2. In parallel with the recursive backstepping design, learning of $h_i(Z_i)$ is also implemented in a recursive procedure. This result can be similarly extended to an nth-order nonlinear strict-feedback system. Note that although learning from direct ANC of system (4.41) appears to be a simple extension of the result in Section 4.2, when considering the indirect ANC approach, learning of system dynamics may not be easy to achieve. This situation is analyzed in the following subsection for a more general class of systems.

4.4   Learning from Direct ANC of Nonlinear Systems in Brunovsky Form

The systems considered in Sections 4.2 and 4.3 have unity control gains that multiply the control term. In this section, we investigate deterministic learning from direct ANC of a more general nonlinear system with unknown affine terms. In many control systems, affine terms often exist in system models (e.g., industrial robots [124]). In the literature of nonlinear control, it is well known that systems with affine terms are more difficult to derive control for and much effort has been devoted to dealing with these terms. From the perspective of learning, the existence of affine terms will also lead to difficulties that prevent accurate parameter convergence (i.e., the occurrence of learning) in the adaptive neural control process. Therefore, to make the deterministic learning control more practical, it is necessary to investigate how to achieve deterministic learning for nonlinear systems in the so-called Brunovsky form [93] with affine terms unknown.

For demonstration of the basic idea, we consider the following second-order nonlinear system in Brunovsky form:

$$\begin{cases} \dot{x}_1 = x_2 \\ \dot{x}_2 = f(x) + g(x)u \end{cases} \tag{4.69}$$

where $x = [x_1, x_2]^T \in R^2$, $u \in R$ are the state variables and system input, respectively, and $f(x)$ and $g(x)$ are unknown smooth nonlinear functions.

As the nature of deterministic learning is related to the exponential stability of a certain class of linear time-varying (LTV) adaptive systems for nonlinear systems in Brunovsky form, the exponential stability of the corresponding LTV adaptive systems will need to be studied first. The difficulty lies in that the unknown affine term $g(x)$ will appear in the closed-loop adaptive system thus causing a special perturbed LTV form. The stability analysis of such LTV systems cannot be handled by existing results of adaptive systems [92,161,173,199]. Another difficulty is that the presence of the affine term $g(x)$ in the closed-loop adaptive system may amplify the NN approximation error and prevent the occurrence of learning even when the exponential stability of the nominal part of the closed-loop adaptive system is achieved. Moreover, the existence of $g(x)$ also leads to more complexity for analyzing the periodicity of NN inputs and the satisfaction of the PE condition.

In this section, we first study the exponential stability of this new class of LTV systems. An extension of the result in [173] is presented which shows that with the satisfaction of a partial PE condition and with some mild conditions, exponential stability of this class of LTV systems can be achieved. Second, to overcome the difficulty caused by the affine term $g(x)$, we introduce a state transformation, by which the closed-loop adaptive system can be turned into the form of perturbed LTV systems with small perturbation terms. Exponential convergence of partial neural weights can be achieved, and deterministic learning from adaptive NN control of nonlinear systems in Brunovsky form can still be implemented. The result will be useful for further research on learning for more general nonlinear systems (such as strict-feedback systems and pure-feedback systems with unknown affine terms [119]), and so be applicable to many industrial applications.

4.4.1 Stability of a Class of Linear Time-Varying Systems

For learning from adaptive NN control of nonlinear systems in Brunovsky form (4.69), the associated LTV system is in the following form:

$$\begin{bmatrix} \dot{e}_1 \\ \dot{e}_2 \\ \dot{\theta} \end{bmatrix} = \begin{bmatrix} & A(t) & \begin{matrix} 0 \\ S^T(t) \end{matrix} \\ \hline 0 \quad -\Gamma S(t)G(t) & 0 \end{bmatrix} \begin{bmatrix} e_1 \\ e_2 \\ \theta \end{bmatrix} \tag{4.70}$$

with $e_1 \in R^{(n-q)}$, $e_2 \in R^q$, $\theta \in R^p$, $A(\cdot) : [0, \infty) \to R^{n \times n}$, $S(\cdot) : [0, \infty) \to R^{p \times q}$, $G(\cdot) : [0, \infty) \to R^{q \times q}$, and $\Gamma = \Gamma^T > 0$. For ease of description, we define

$$e := \begin{bmatrix} e_1^T & e_2^T \end{bmatrix}^T \in R^n \tag{4.71}$$

$$\eta := \begin{bmatrix} e^T & \theta^T \end{bmatrix}^T \in R^{(n+p)} \tag{4.72}$$

$$B(t) := \begin{bmatrix} 0 & S(t) \end{bmatrix} \in R^{p \times n} \tag{4.73}$$

$$P(t) := diag\{I, G(t)\} \in R^{n \times n} \tag{4.74}$$

$$C(t) := \begin{bmatrix} 0 & \Gamma S(t) G(t) \end{bmatrix} \in R^{p \times n} \tag{4.75}$$

where *diag* here refers to block diagonal form.

It follows that

$$C(t) = \Gamma B(t) P(t) \tag{4.76}$$

There is no specific result for exponential stability of system (4.70). Existing results on LTV systems (e.g., Theorems 2.4 and 2.5) are useful, but they cannot be applied directly for stability analysis of system (4.70). In Theorem 2.4, the matrix A in system (2.18) is time-invariant, whereas the matrix $A(t)$ in system (4.70) is time-varying. On the other hand, although the LTV system (2.19) considered in Theorem 2.5 contains a time-varying matrix $A(t)$, we still cannot apply Theorem 2.5 directly because $B(t) = \begin{bmatrix} 0 & S(t) \end{bmatrix}^T$ in system (4.70) implies that PE of $B(t)$ cannot be satisfied.

Based on Theorems 2.4 and 2.5, we give the following lemma on the exponential stability of system (4.70), in which $B(t) = \begin{bmatrix} 0 & S(t) \end{bmatrix}^T$ does not satisfy the PE condition. We introduce a weaker version of Assumption 2.3.

ASSUMPTION 4.2
There exist symmetric matrices $P(t)$ and $Q(t)$ such that $A^T(t)P(t) + P(t)A(t) + \dot{P}(t) = -Q(t)$. Furthermore, $\exists$ p_m, q_m, p_M, and $q_M > 0$ such that $p_m I \leq P(t) \leq p_M I$ and $q_m I \leq Q(t) \leq q_M I$.

LEMMA 4.1
The system (4.70) with Assumptions 2.1 and 2.2 and Assumption 4.2 satisfied in a compact set Ω is uniformly exponentially stable in Ω if $S(t)$ satisfies the PE condition.

PROOF   Our proof is motivated by the proof of Theorem 2.5 given in [173].
Consider the Lyapunov function candidate

$$V_1 = \frac{1}{2}e^T P(t)e + \frac{1}{2}\theta^T \Gamma^{-1}\theta \tag{4.77}$$

Then, the derivative of V_1 is

$$\dot{V}_1 = \frac{1}{2}e^T P\dot{e} + \frac{1}{2}\dot{e}^T Pe + \frac{1}{2}e^T \dot{P}e + \theta^T \Gamma^{-1}\dot{\theta}$$

$$= \frac{1}{2}e^T(PA + A^T P + \dot{P})e$$

$$= -\frac{1}{2}e^T Q(t)e \le -\frac{1}{2}q_m \|e\|^2 . \tag{4.78}$$

Thus, system (4.70) is uniformly stable.

Let $a > 0$, and define

$$\Xi(t) := \begin{bmatrix} \bar{A} & B^T(t) \\ -\Gamma B(t) & 0 \end{bmatrix} \tag{4.79}$$

$$\Psi(t, e) := \begin{bmatrix} [A(t) - \bar{A}]e \\ \Gamma B(t)[I - P(t)]e \end{bmatrix} \tag{4.80}$$

where $\bar{A} = -a I$; then system (4.70) can be rewritten as

$$\dot{\eta} = \Xi(t)\eta + \Psi(t, e) \tag{4.81}$$

From Assumptions 2.1, 2.2, and 4.2, there exists a $k_g > 0$, such that $\|\Psi(t, e)\| \le k_g \|e\|$. From Theorem 2.4, when $S(t)$ satisfies the PE condition, the system $\dot{\eta} = \Xi(t)\eta$ is exponentially stable. From Theorem 4.12 in [111], there exists a Lyapunov function

$$V_2 = \eta^T P_0(t)\eta \tag{4.82}$$

for $\dot{\eta} = \Xi(t)\eta$, such that V_2 satisfies

$$c_1 \|\eta\|^2 \le V_2 \le c_2 \|\eta\|^2 \tag{4.83}$$

$$\dot{V}_2 \le -c_3 \|\eta\|^2 \tag{4.84}$$

Along the trajectory of system (4.70), the derivative of V_2 satisfies

$$\dot{V}_2 = \eta^T P_0\dot{\eta} + \dot{\eta}^T P_0\eta + \eta^T \dot{P}_0\eta$$

$$= \eta^T P_0(t)\Xi(t)\eta + \eta^T \Xi^T(t)P_0(t)\eta + \eta^T \dot{P}_0\eta + 2\eta^T P_0\Psi(t)$$

$$< -c_3 \|\eta\|^2 + 2c_2 k_g \|e\| \|\eta\| \tag{4.85}$$

For system (4.70), we define the following Lyapunov function candidate

$$V_3 = \pi V_1 + V_2 \tag{4.86}$$

with π a positive constant.

Then, the derivative of V_3 satisfies

$$\dot{V}_3 < -\pi q_m \|e\|^2 - c_3 \|\eta\|^2 + 2c_2 k_g \|e\| \|\eta\| \tag{4.87}$$

If we choose

$$\pi \geq \frac{2k_g^2 c_2^2}{q_m c_3}$$

then

$$\dot{V}_3 \leq -\frac{c_3}{2}\|\eta\|^2 \tag{4.88}$$

This ends the proof. ■

Lemma 4.1 implies that for system (4.70), even though $B(t) = [0 \quad S(t)]^T$ cannot satisfy the PE condition, the PE of $S(t)$ can still lead to the exponential stability of the LTV system. On the other hand, to use Lemma 4.1, it is necessary to transform the adaptive NN control system into a perturbed LTV system with a small perturbation term.

4.4.2 Learning from Direct ANC

For nonlinear systems in Brunovsky form (4.69), we make the following assumptions.

ASSUMPTION 4.3
The sign of $g(x)$ is known, and there exist constants $g_1 \geq g_0 > 0$ such that $g_1 \geq |g(\cdot)| \geq g_0$, $\forall x \in \Omega \subset R^2$. Without losing generality, we assume $g_1 \geq g(x) \geq g_0$, $\forall x \in \Omega \subset R^2$.

ASSUMPTION 4.4
There exists a constant $g_d > 0$ such that $|\dot{g}(x)| \leq g_d$, $\forall x \in \Omega \subset R^2$, where the derivative is with respect to time.

The reference model is the same system expressed by Equation (4.2) with Assumption 4.1:

$$\begin{cases} \dot{x}_{d_1} = x_{d_2} \\ \dot{x}_{d_2} = f_d(x_d) \end{cases} \tag{4.89}$$

Our objective is to develop an ANC using localized RBF networks such that (i) all the signals in the closed-loop system are uniformly bounded, and (ii) accurate NN approximation (learning) of the closed-loop control system dynamics can be achieved in a local region along an orbit of recurrent closed-loop signals as previously achieved in Theorems 4.1 and 4.2.

For system (4.69) and reference model (4.89), an ANC similar to one in [65]) is designed using a Gaussian RBFN as follows:

$$u = -z_1 - c_2 z_2 - \widehat{W}^T S(Z) \tag{4.90}$$

where

$$z_1 = x_1 - x_{d_1} \tag{4.91}$$

$$z_2 = x_2 - \alpha_1 \tag{4.92}$$

$$\alpha_1 = -c_1 z_1 + \dot{x}_{d_1} = -c_1 z_1 + x_{d2} \tag{4.93}$$

$$\dot{\alpha}_1 = -c_1 \dot{z}_1 + \dot{x}_{d_2} = -c_1(-c_1 z_1 + z_2) + f_d(x_d) \tag{4.94}$$

and $c_1, c_2 > 0$ are control gains. The Gaussian RBFN $\widehat{W}^T S(Z)$ is used to approximate the unknown function

$$h(Z) = (f(x) - \dot{\alpha}_1)/g(x) \tag{4.95}$$

where $Z = [x_1, x_2, \dot{\alpha}_1]^T \in \Omega \subset R^3$ is the NN input, and $\widehat{W}$ is the estimate of its optimal value W^*, and is updated by

$$\dot{\widehat{W}} = \dot{\widetilde{W}} = \Gamma\left(S(Z)z_2 - \sigma \widehat{W}\right) \tag{4.96}$$

where $\widetilde{W} = \widehat{W} - W^*$, and $\Gamma = \Gamma^T > 0$ is a design matrix in diagonal form.

REMARK 4.10
The controller design [64, Section 7.2] uses the controller function $h(Z)$ to achieve partial feedback linearization. Note, however, that Equation (4.95) does not reduce to the unknown function in the $g(x) = 1$ case—see Section 4.2—due to the presence of $\dot{\alpha}_1$ as the input to the NN.

The overall closed-loop system can be summarized in the following form:

$$\begin{cases} \dot{z}_1 = -c_1 z_1 + z_2 \\ \dot{z}_2 = -g(x)\left[z_1 + c_2 z_2 + \widetilde{W}^T S(Z) - \epsilon(Z)\right] \\ \dot{\widehat{W}} = \dot{\widetilde{W}} = \Gamma\left(S(Z)z_2 - \sigma \widehat{W}\right) \end{cases} \tag{4.97}$$

which has a similar form to Equation (4.12), except that the affine term $g(x)$ now appears in Equation (4.97).

THEOREM 4.4 **(Stability and Tracking)**
Consider the closed-loop system (4.97) consisting of the plant (4.69), the reference model (4.89), the controller (4.90), and the NN adaptation law (4.96). For a sufficiently large compact set Ω, with initial conditions appropriately chosen, and with $\widehat{W}(0) = 0$, we have that: (i) all the signals in the closed-loop system remain uniformly bounded; (ii) there exists a time Υ_1 such that the NN input $Z = [x_1, x_2, \dot{\alpha}_1]^T$ converges to a small neighborhood of periodic signal $Z_d(t) = [x_{d_1}(t), x_{d_2}(t), f_d(x_d(t))]^T$ for all $t \geq \Upsilon_1$ by appropriately choosing design parameters.

PROOF (i) Boundedness of all signals in the closed-loop can be proved similarly to [65]. The details are omitted here.

(ii) To achieve objective (ii), we require that without the PE condition, x converges arbitrarily close to x_d in a finite time Υ_1.

Following the analysis of adaptive neural control (see Section 4.2 for details), by appropriately choosing the controller parameters, there exist a small constant μ and a finite time Υ_1, such that both z_1 and z_2 satisfy

$$|z_i(t)| < \mu, \quad i = 1, 2 \tag{4.98}$$

Since $z_1 = x_1 - x_{d_1}$, we know that x_1 will converge to x_{d_1}. From $z_2 = x_2 - \alpha_1 = x_2 + c_1 z_1 - x_{d_2}$, we get

$$|x_2 - x_{d_2}| = |z_2 - c_1 z_1| \leq |z_2| + c_1|z_1| \leq (1 + c_1)\mu \tag{4.99}$$

which is a small value when μ is small. Because $\dot{\alpha} - f_d(x_d) = -c_1(-c_1 z_1 + z_2)$, we have

$$\begin{aligned}
|\dot{\alpha}_1 - f_d(x_d)| &= |-c_1(-c_1 z_1 + z_2)| \\
&\leq c_1^2|z_1| + c_1|z_2| \\
&\leq c_1(1 + c_1)\mu
\end{aligned} \tag{4.100}$$

which is also small when μ is small, and c_1 is appropriately chosen.

Thus, x_1, x_2, and $\dot{\alpha}_1$ converge closely to x_{d_1}, x_{d_2}, and $f_d(x_d)$ in finite time Υ_1. Therefore, the NN input $Z = [x_1, x_2, \dot{\alpha}_1]^T$ is made as recurrent as $Z_d = [x_{d_1}, x_{d_2}, f_d(x_d)]^T$ for all $t \geq \Upsilon_1$. This ends the proof. ∎

To achieve deterministic learning for closed-loop system (4.97), two difficulties arise: (i) the satisfaction of the PE condition of $S(Z)$; and (ii) exponential stability of the closed-loop control system. In Sections 4.2 and 4.3, the first difficulty has been successfully overcome in two steps: (1) state tracking convergence in finite time by adaptive neural control without the PE condition, and (2) satisfaction of the PE condition for a regression subvector $S_\zeta(Z)$ thanks to the properties of RBF networks and the state tracking.

For the second difficulty, because of the existence of affine term $g(x)$, exponential stability of closed-loop control system (4.97) cannot be guaranteed directly using existing stability theorems of adaptive control [92,161,199]. Compared with the results discussed in Section 4.2, (4.97) represents a more general adaptive system. To overcome this difficulty, we introduce a state transformation, such that system (4.97) is described in the form of LTV system (4.70), and exponential stability of the closed-loop system is achieved by using Lemma 4.1.

By using the local property of the Gaussian RBF network, after time Υ_1, system (4.97) can be expressed in the following form along the tracking orbit

$\varphi_\zeta(Z(t))|_{t \geq \Upsilon}$ as:

$$\begin{cases} \dot{z}_1 = -c_1 z_1 + z_2 \\ \dot{z}_2 = -g(x)\left[z_1 + c_2 z_2 + \widetilde{W}_\zeta^T S_\zeta(Z) - \epsilon'_\zeta\right] \\ \dot{\widehat{W}}_\zeta = \dot{\widetilde{W}}_\zeta = \Gamma_\zeta(S_\zeta(Z)z_2 - \sigma \widehat{W}_\zeta) \end{cases} \tag{4.101}$$

$$\dot{\widehat{W}}_{\bar{\zeta}} = \dot{\widetilde{W}}_{\bar{\zeta}} = \Gamma_{\bar{\zeta}}(S_{\bar{\zeta}}(Z)z_2 - \sigma \widehat{W}_{\bar{\zeta}}) \tag{4.102}$$

where $S_\zeta(Z)$ is a subvector of $S(Z)$, $\widehat{W}_\zeta$ is the corresponding weight subvector, the subscript $\bar{\zeta}$ stands for the region far away from the trajectory $\varphi_\zeta(Z(t))|_{t \geq \Upsilon_1}$, and $\epsilon'_\zeta = \epsilon_\zeta - \widetilde{W}_{\bar{\zeta}}^T S_{\bar{\zeta}}(z) = O(\epsilon_\zeta)$.

THEOREM 4.5 (**Learning**)
Consider the closed-loop system (4.97) consisting of the plant (4.69), the reference model (4.89), the controller (4.90), and the NN adaptation law (4.96). For a sufficiently large compact set Ω, with initial conditions and control parameters appropriately chosen, and with $\widehat{W}(0) = 0$, we have that the neural-weight estimates $\widehat{W}_\zeta$ converge to small neighborhoods of their optimal values W_ζ^*, and the locally accurate approximation of controller dynamics $h(Z) = (f(x) - \dot{\alpha}_1)/g(x)$ along the tracking orbit $\varphi_\zeta(z(t))|_{t \geq \Upsilon}$ is obtained by $\widehat{W}^T S(Z)$ to the error level ϵ, as well as by $\overline{W}^T S(Z)$, where

$$\overline{W} = mean_{t \in [t_a, t_b]} \widehat{W}(t) \tag{4.103}$$

with $[t_a, t_b]$, $t_b > t_a > \Upsilon$ representing a time segment after the transient process.

PROOF The closed-loop system can be represented in the following LTV form:

$$\begin{bmatrix} \dot{z}_1 \\ \dot{z}_2 \\ \dot{\widetilde{W}}_\zeta \end{bmatrix} = \begin{bmatrix} -c_1 & 1 & 0 \\ -g(x) & -c_2 g(x) & -g(x)S_\zeta^T(Z) \\ 0 & \Gamma_\zeta S_\zeta(Z) & 0 \end{bmatrix} \begin{bmatrix} z_1 \\ z_2 \\ \widetilde{W}_\zeta \end{bmatrix}$$

$$+ \begin{bmatrix} 0 \\ g(x)\epsilon'_\zeta \\ -\sigma \Gamma_\zeta \widehat{W}_\zeta \end{bmatrix} \tag{4.104}$$

We introduce a state transformation to modify the influence of the perturbation term caused by the NN approximation error. Then the parameter convergence can be guaranteed by exponential stability of the nominal system.

Let $e_1 = z_1$, $e_2 = z_2/g(x)$, and $\theta = \widetilde{W}_\zeta$ (with a little abuse of notation), then system (4.101) is transformed into

$$\begin{cases} \dot{e}_1 = -c_1 e_1 + g(x)e_2 \\ \dot{e}_2 = -e_1 - \left[c_2 g(x) + \frac{\dot{g}(\cdot)}{g(x)}\right]e_2 - \theta^T S_\zeta^T(Z) + \epsilon'_\zeta \\ \dot{\theta} = \Gamma_\zeta g(x)S_\zeta(Z)e_2 - \sigma \Gamma_\zeta \widehat{W}_\zeta \end{cases} \tag{4.105}$$

that is,

$$
\begin{bmatrix} \dot{e}_1 \\ \dot{e}_2 \\ \dot{\theta} \end{bmatrix} = \left[\begin{array}{cc|cc} & A(t) & 0 & -S_\zeta^T(Z) \\ \hline 0 & \Gamma_\zeta g(x) S_\zeta(Z) & 0 & \end{array} \right] \begin{bmatrix} e_1 \\ e_2 \\ \theta \end{bmatrix}
$$
$$
+ \begin{bmatrix} 0 \\ \epsilon_\zeta' \\ -\sigma \Gamma_\zeta \widehat{W}_\zeta \end{bmatrix} \tag{4.106}
$$

with

$$
A(t) = \begin{bmatrix} -c_1 & g(x) \\ -1 & -\left[c_2 g(x) + \frac{\dot{g}(t)}{g(x)} \right] \end{bmatrix} \tag{4.107}
$$

Because $|\epsilon_\zeta'|$ and $\|\sigma \Gamma_\zeta \widehat{W}_\zeta\|$ are small, system (4.106) can be considered a perturbed system [111].

Consider the nominal part of perturbed system (4.106); that is,

$$
\begin{bmatrix} \dot{e}_1 \\ \dot{e}_2 \\ \dot{\theta} \end{bmatrix} = \left[\begin{array}{cc|cc} & A(t) & 0 & -S_\zeta^T(Z) \\ \hline 0 & \Gamma_\zeta S_\zeta(Z) g(x) & 0 & \end{array} \right] \begin{bmatrix} e_1 \\ e_2 \\ \theta \end{bmatrix} \tag{4.108}
$$

Let

$$
B^T(t) = - \begin{bmatrix} 0 \\ S_\zeta^T(Z(t)) \end{bmatrix} \tag{4.109}
$$

$$
P(t) = \begin{bmatrix} 1 & 0 \\ 0 & g(x(t)) \end{bmatrix} \tag{4.110}
$$

Then from the definitions of $A(t)$ and $P(t)$ in (4.107) and (4.110), we have

$$
\dot{P} + PA + A^T P = \begin{bmatrix} -2c_1 & 0 \\ 0 & -2c_2 g^2(x) - \dot{g}(x) \end{bmatrix} \tag{4.111}
$$

The satisfaction of Assumption 2.1 can be easily checked. From Assumptions 4.3 and 4.4, c_2 can be found such that

$$
2c_2 + \frac{\dot{g}(x)}{g^2(x)} > 0
$$

and the negative definiteness of $\dot{P} + PA + A^T P$ is guaranteed with P positive definite. Thus, Assumption 4.2 is satisfied.

From Theorem 4.4, after time Υ_1, the NN input can follow a recurrent orbit, and the partial PE condition [242] can be satisfied by the regression subvector $S_\zeta(Z)$, which consists of RBFs with centers located in a neighborhood of the tracking orbit $\varphi_\zeta(Z(t))|_{t\geq\Upsilon_1}$.

Then, for the nominal system (4.108), uniform exponential stability is guaranteed by Lemma 4.1. For the perturbed system (4.106), by using Theorem 2.6, the parameter error $\theta = \widetilde{W}_\zeta$ converges exponentially to a small neighborhood of zero in a finite time Υ, with the sizes of the neighborhoods being determined by ϵ^* and $\sigma\|\Gamma_\zeta\widehat{W}_\zeta^*\|$.

The convergence of $\widehat{W}_\zeta$ to a small neighborhood of W_ζ^* implies that along the trajectory $\varphi_\zeta(Z(t))|_{t\geq\Upsilon}$, we have

$$h(Z) = W_\zeta^{*T}S_\zeta(Z) + \epsilon_\zeta$$
$$= \widehat{W}_\zeta^T S_\zeta(Z) - \widetilde{W}_\zeta^T S_\zeta(Z) + \epsilon_\zeta$$
$$= \widehat{W}_\zeta^T S_\zeta(Z) + \epsilon_{\zeta_1} \tag{4.112}$$

where $\epsilon_{\zeta_1} = \epsilon_\zeta - \widetilde{W}_\zeta^T S_\zeta(Z) = 0(\epsilon_\zeta)$ is close to ϵ_ζ due to the convergence of $\widetilde{W}_\zeta^T S_\zeta(Z)$.

Choosing $\overline{W}$ according to Equations (4.103) and (4.112) can be expressed as

$$h(Z) = \widehat{W}_\zeta^T S_\zeta(Z) + \epsilon_{\zeta_1}$$
$$= \overline{W}_\zeta^T S_\zeta(Z) + \epsilon_{\zeta_2} \tag{4.113}$$

where $\overline{W}_\zeta^T = [\overline{W}_{j_1}, \ldots, \overline{W}_{j_\zeta}]^T$ is the subvector of $\overline{W}$, and ϵ_{ζ_2} is an error arising from using $\overline{W}_\zeta^T S_\zeta(Z)$ as the system approximation. It is clear that after the transient process, $\epsilon_{\zeta_2} = O(\epsilon_{\zeta_1})$.

On the other hand, due to the localization property of Gaussian RBFs, both $S_{\bar{\zeta}}(Z)$ and $\overline{W}_{\bar{\zeta}}^T S_{\bar{\zeta}}(Z)$ are very small. This means that along trajectory $\varphi_\zeta(z(t))|_{t\geq\Upsilon}$ the entire RBF network $\widehat{W}^T S(Z)$ and $\overline{W}^T S(Z)$ can approximate the unknown $h(Z)$ as

$$h(Z) = W_\zeta^{*T}S_\zeta(Z) + \epsilon_\zeta$$
$$= \widehat{W}^T S(Z) + \epsilon_1$$
$$= \overline{W}^T S(Z) + \epsilon_2 \tag{4.114}$$

where $\epsilon_1 = O(\epsilon_\zeta) = O(\epsilon)$, $\epsilon_2 = O(\epsilon_{\zeta_2}) = O(\epsilon)$. It is seen that $\widehat{W}^T S(Z)$ and $\overline{W}^T S(Z)$ are capable of approximating the unknown nonlinearity $h(Z)$ along the tracking orbit $\varphi_\zeta(Z(t))|_{t\geq\Upsilon}$ [and the reference orbit $\varphi_d(Z_d(t))$] to the error level ϵ. This ends the proof. ∎

REMARK 4.11

In the above analysis, it is seen that for nonlinear systems in Brunovsky form, closed-loop identification of $h(Z)$ is achieved. Note that the closed-loop dynamics $h(Z)$ is not simply a nonlinear function of the plant, but the control system dynamics determined by the plant, the reference model, and the controller. Thus, from the viewpoint of system identification, deterministic learning provides a simple and effective approach for identification of closed-loop dynamics.

REMARK 4.12

For indirect adaptive NN control, in which neural networks are used to approximate the system dynamics of the plant, for example, $f(x)$ and $g(x)$ in Equation (4.69), or $f_1(x_1)$ and $f_2(x_1, x_2)$ in Equation (4.41), the stability proof tends to be much more algebraically involved than the proof of the direct ANC approach [269]. Concerning the issues of learning from indirect ANC, it is analyzed in [240] that the indirect ANC approach may not lead to accurate approximations of system dynamics $f(x)$ and $g(x)$ in Equation (4.69) even when the PE condition is satisfied. From the perspective of learning, it appears that the direct ANC approach is simpler to guarantee learning than the indirect approach. Although there are difficulties in establishing learning from the indirect ANC of nonlinear systems, a detailed comparison requires more study.

4.4.3   Simulation Studies

To verify the neural learning and control approach presented in this section, the following plant is taken:

$$\dot{x}_1 = x_2$$
$$\dot{x}_2 = -x_1 + 0.7(1 - x_1^2)x_2 + (2 + 0.5\sin x_1)u \qquad (4.115)$$

where the smooth functions $f(x_1, x_2) = -x_1 + 0.7(1 - x_1^2)x_2$ and $g(x_1) = 2 + 0.5\sin x_1$ are considered as unknown in the controller design.

The reference trajectory is generated from the Duffing oscillator (4.40), with parameters $p_1 = 0.4$, $p_2 = -1.1$, $p_3 = 1.0$, $w = 1.8$, and $q = 1.498$. The initial states of the reference model are $[x_{d_1}(0), x_{d_2}(0)]^T = [0.2, 0.3]^T$ as shown in Figure 4.5.

We construct the Gaussian RBF network $W^T S(Z)$ using 243 nodes (i.e., $N = 243$), with the centers μ_i evenly spaced on $[-3.0, 3.0] \times [-3.0, 3.0] \times [-3.0, 3.0]$, and the widths $\eta_i = 1.5$. The design parameters are $c_1 = 10$, $c_2 = 15$, $\Gamma = 10$, and $\sigma = 0.01$. The initial weights $\widehat{W}(0) = 0$, and the initial states $[x_1(0), x_2(0)]^T = [0, 0]^T$.

The state tracking performance is shown in Figure 4.5. The control input is shown in Figure 4.6. In Figure 4.7, the parameter convergence is shown, and it is clear that the L_2 norm of the NN weights W converges to a value. From Figure 4.8, it can be seen more intuitively that just part of the neural

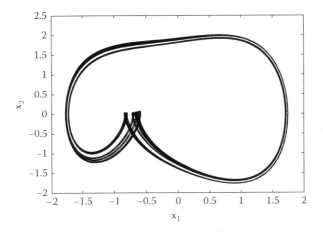

FIGURE 4.5
Tracking convergence: x ("$-$"), x_d ("$--$").

weights converge to relatively larger values, while many other neural weights remain 0 or a small values. This is consistent with satisfaction of the partial PE condition. Figure 4.9 shows the approximation of the control system dynamics $h(Z)$.

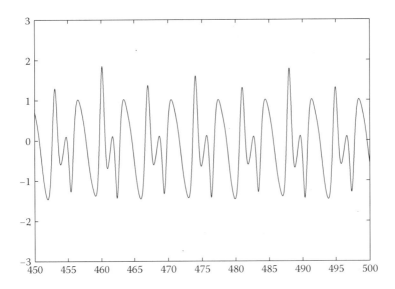

FIGURE 4.6
The control u of adaptive neural control.

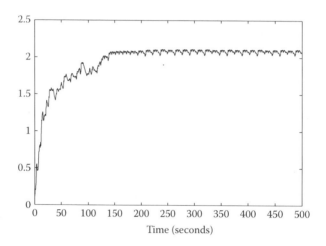

FIGURE 4.7
Parameter convergence $\|\widehat{W}\|$.

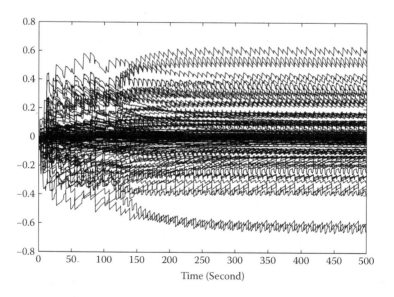

FIGURE 4.8
Partial parameter convergence $\widehat{W}$.

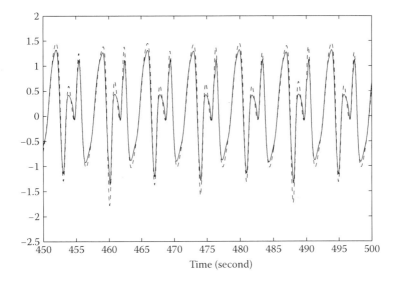

FIGURE 4.9
Function approximation: $h(Z)("-")$, $h_{nn}(Z)("--")$.

4.5 Summary

As control is often the main motivation for system identification in the systems and control community, identification for model-based control has led to a challenging problem of closed-loop identification [48]. The basic idea of closed-loop identification is that the estimated models are acceptable as long as the control performance is achieved [48]. In other words, because identification of a true nonlinear system model is too difficult to be achieved, identification of a true closed-loop system model is not the objective in the literature of system identification, and so is considered unnecessary.

In this chapter, we have presented methods for deterministic learning from closed-loop control of several classes of nonlinear systems. It has been shown that locally accurate closed-loop identification of the unknown system dynamics can be achieved during tracking control to recurrent reference orbits via direct adaptive NN control. Specifically, the partial PE condition of the internal closed-loop signals has been shown to be satisfied when the system states closely track the recurrent states of the reference model, and locally accurate NN approximation of closed-loop system dynamics is achieved in a region along the recurrent tracking orbit. For the neurons centered close to the tracking orbit, their neural weights converge to a small neighborhood of a set of optimal values, while for the other neurons far away from the tracking orbit, the neural weights are updated only slightly. Thus, it has been shown that deterministic learning is capable of obtaining knowledge of control system dynamics from closed-loop control processes. The knowledge obtained can be utilized in another similar control task to achieve guaranteed stability and improved control performance. As we will see later, the capabilities of deterministic learning control systems for obtaining and utilizing knowledge reveal a higher level of intelligence and a higher degree of autonomy compared with conventional adaptive control systems.

5

Dynamical Pattern Recognition

5.1 Introduction

Recognition of temporal or dynamical patterns is among the most difficult tasks in the pattern recognition area. Nonetheless, it is noticed that humans generally excel in dealing with such patterns as they do in speech recognition, high performance sports, and rescue operations. Human recognition of temporal patterns is an integrated process, in which patterns of information distributed over time can be effectively identified, represented, recognized, and classified. A distinguishing feature of the human recognition process is that it takes place quickly from the beginning of sensing temporal patterns, and runs directly on the input space for feature extraction and pattern matching. These recognition mechanisms, although not fully understood, appear to be quite different from the existing neural network and statistical approaches for pattern recognition. Although a great deal of progress has been made in the area of recognition of static patterns, only limited success has been reported in the literature for rapid recognition of temporal patterns.

One early result for classification of spatio-temporal patterns is Grossberg's formal avalanche structure [72]. A popular approach for temporal pattern processing is to construct short-term memory (STM) models, such as delay lines [236], decay traces [101,251], and exponential kernels [217]. These STM models are then embedded into different neural network architectures. For example, the time delay neural network (TDNN) is proposed by combining multilayer perceptrons (MLPs) with the delay line model [236]. With STM models, a temporal pattern is represented as a sequence of pattern states, and recognition of temporal patterns is made quite similar to the recognition of static patterns. From our point of view, it appears a limited approach to treat temporal patterns as multiple static patterns.

In temporal pattern recognition, there are some fundamental issues that need to be addressed. Among the numerous unresolved problems in this field, one of the most fundamental issues is how to appropriately represent the time-varying patterns in a time-independent manner [34]. Another important problem currently studied in this area is the definition of similarity between two temporal patterns. As temporal patterns evolve with time, the existing similarity measures developed for static patterns do not seem appropriate.

In this chapter, we investigate the recognition of a class of temporal patterns, which are generated from a general nonlinear dynamical system:

$$\dot{x} = F(x; p), \qquad x(t_0) = x_0 \tag{5.1}$$

where $x = [x_1, \ldots, x_n]^T \in R^n$ is the state of the system, p is a vector of system parameters, and $F(x; p) = [f_1(x; p), \ldots, f_n(x; p)]^T$ represents the system dynamics, in which each $f_i(x; p)$ is an unknown, continuous nonlinear function.

A *dynamical pattern* is defined as a recurrent system trajectory generated from the above dynamical system. The class of recurrent trajectories includes periodic, quasi-periodic, almost-periodic, and even chaotic trajectories, which are some of the most important types of trajectories generated from nonlinear dynamical systems. Nonlinear dynamical system theory has been found useful for explanation of the formation of numerous dynamical patterns in areas such as hydrodynamics, oceanography, meteorology, biological morphodynamics, and semiconductors [14,75,187]. In other words, nonlinear dynamical systems are capable of exhibiting various types of dynamical patterns. Therefore, the definition of a dynamical pattern above covers a wide class of temporal patterns studied in the literature.

The general recognition process for a dynamical pattern usually consists of two phases: the identification phase and the recognition phase. Here, "identification" involves working out the essential features of a pattern one does not recognize, whereas "recognition" means looking at a pattern and realizing that it is the same or a similar pattern to one seen earlier. For identification of dynamical patterns, we can use deterministic learning for nonlinear dynamical systems as described in Chapter 3. Locally accurate NN approximation of the underlying system dynamics $F(x; p)$ within a dynamical pattern can be achieved by using localized RBF networks. Through deterministic learning, fundamental knowledge of dynamical patterns is obtained in the identification phase and is stored as constant RBF neural weights.

In this chapter, based on the deterministic learning mechanism presented in Chapter 3, a unified framework is proposed for effective representation, similarity characterization, and rapid recognition of dynamical patterns. First in Section 5.2, it is shown that a time-varying dynamical pattern can be effectively represented in a time-invariant and spatially distributed manner through deterministic learning. Second, a definition for characterizing similarity of dynamical patterns is given in Section 5.3 based on system dynamics inherently within dynamical patterns. Third, in Section 5.4, a mechanism for rapid recognition of dynamical patterns is presented, which reveals how the learned knowledge is utilized in the recognition phase. A test dynamical pattern is recognized as similar to a training dynamical pattern if state synchronization is achieved according to a kind of internal and dynamical matching on system dynamics. The synchronization errors can be taken as the measure of similarity between the test and training patterns. It is shown that due to knowledge utilization, the problem of dynamical pattern recognition is converted into one of the stability and convergence of a linear time-invariant

(LTI) recognition error system. Finally in Section 5.5, the construction of recognition systems for dynamical pattern classification is investigated. The work of this chapter draws substantially on the papers [239,244].

5.2 Time-Invariant Representation

In static pattern recognition, a pattern is usually a set of time-invariant measurements or observations represented in vector or matrix notation [19,95]. The dimensionality of the vector or matrix representation is generally kept as small as possible by using a limited yet salient feature set for purposes such as removing redundant information and improving classification performance. For example, in statistical pattern recognition, a pattern is represented by a set of d features, or a d-dimensional feature vector which yields a d-dimensional feature space. Subsequently, the task of recognition or classification is accomplished when the d-dimensional feature space is partitioned into compact and disjoint regions, and decision boundaries are constructed in the feature space that separate patterns from different classes into different regions [95,254].

For dynamical patterns, because the measurements are mostly time-varying in nature, the above framework for static patterns may not be suitable for representation of dynamical patterns. As indicated in [34], if the time attribute could not be appropriately dealt with, the problem of *time-independent* representation without loss of discrimination power and classification accuracy would be a very difficult task for temporal/dynamical pattern recognition. Furthermore, without a proper representation of dynamical patterns, the problem of how to define the similarity between two dynamical patterns will become another difficulty.

In this section, based on deterministic learning theory, we show that by using the constant RBF networks obtained through deterministic learning, time-varying dynamical patterns can be effectively represented by the locally accurate NN approximations of system dynamics $F(x; p)$. The information is stored by a large number of neurons distributed along the state trajectory of a dynamical pattern. It is shown that the representation is essential for similarity definition and rapid recognition of dynamical patterns.

5.2.1 Static Representation

As introduced in Chapter 3, the system dynamics $F(x; p) = [f_1(x; p), \ldots, f_n(x; p)]^T$ of a dynamical pattern φ_ζ can be accurately approximated by $\overline{W}_i^T S(x)$ $(i = 1, \ldots, n)$ in a local region along the recurrent orbit of the dynamical pattern φ_ζ. The *constant* RBF network $\overline{W}_i^T S(x)$ consists of two types of neural weights: (i) for neurons whose centers are close to the orbit $\varphi_\zeta(x_0)$, their neural weights $\widehat{W}_{\zeta i}$ converge exponentially to a small neighborhood of their optimal values $W_{\zeta i}^*$; and (ii) for the neurons with centers far away from the orbit $\varphi_\zeta(x_0)$, the neural weights $\widehat{W}_{\bar{\zeta} i}$ will remain almost 0. Thus, constant neural weights are obtained for all neurons of the entire RBF network $\overline{W}_i^T S(x)$. Accordingly, from Theorem 3.1 and Corollary 3.1, we have the following statements concerning the representation of a dynamical pattern:

1. A dynamical pattern φ_ζ can be represented by using the constant RBF network $\overline{W}_i^T S_i(x)$ ($i = 1, \ldots, n$), which provides an NN approximation of the *time-invariant* system dynamics $f_i(x; p)$ ($i = 1, \ldots, n$). This representation, based on the fundamental information extracted from the dynamical pattern φ_ζ, is independent of time. The NN approximation $\overline{W}_i^T S_i(x)$ is accurate only in a local region (denoted as Ω_{φ_ζ}) along the orbit $\varphi_\zeta(x_0)$. The locally accurate NN approximation provides an efficient solution to the problem of representation of time-varying dynamical patterns.

2. The representation by $\overline{W}_i^T S_i(x)$ is *spatially distributed* in the sense that relevant information is stored in a large number of neurons distributed along the state trajectory of a dynamical pattern. It shows that for appropriate representation of a dynamical pattern, complete information on both the pattern state and the underlying system dynamics is utilized. Specifically, a dynamical pattern is represented by using information on its state trajectory (starting from an initial condition), plus its underlying system dynamics along the state trajectory. Intuitively, the spatially distributed information implies that a representation using a limited number of extracted features (as in static pattern recognition) is probably incomplete for representation of dynamical patterns in many situations.

Concerning the locally accurate NN approximation, the local region Ω_{φ_ζ} is described by

$$\Omega_{\varphi_\zeta} := \left\{ x \mid \text{dist}(x, \varphi_\zeta) < d \Rightarrow \left| \overline{W}_i^T S_i(x) - f_i(x; p) \right| < \xi_i^*, i = 1, \ldots, n \right\} \quad (5.2)$$

where $d, \xi_i^* > 0$ are constants; $\xi_i^* = O(\epsilon_i^*)$ is the approximation error within Ω_{φ_ζ}. This knowledge stored in $\overline{W}_i^T S_i(x)$ can be recalled in such a way that whenever the NN input $Z(= x)$ enters the region Ω_{φ_ζ}, the RBF network $\overline{W}_i^T S_i(x)$ will provide accurate approximation to the dynamics $f_i(x; p)$.

5.2.2   Dynamic Representation

Note that the representation by $\overline{W}_i^T S_i(x)$ is not used directly for recognition, that is, recognition by direct comparison of the corresponding neural weights. Instead, for a training dynamical pattern φ_ζ, we construct a dynamical model using $\overline{W}_i^T S_i(x)$ ($i = 1, \ldots, n$) as:

$$\dot{\bar{x}} = -B(\bar{x} - x) + \overline{W}^T S_A(x) \quad (5.3)$$

where $\bar{x} = [\bar{x}_1, \ldots, \bar{x}_n]^T$ is the state of the dynamical model, x is the state of an input pattern generated from system (5.1), $\overline{W}^T S_A(x) = [\overline{W}_1^T S_1(x), \ldots, \overline{W}_n^T S_n(x)]^T$ are constant RBF networks obtained through deterministic learning, $B = diag\{b_1, \ldots, b_n\}$ is a diagonal matrix, with $b_i > 0$ normally smaller than a_i (a_i is given in Equation [3.2]) and $S_A(x) = diag\{S_1(x), \ldots, S_n(x)\}$.

It is clearly seen that the representation of dynamical patterns is quite different from the representation used in static pattern recognition. As detailed in Section 5.4, the dynamical model (5.3) is used as a representative of the training dynamical pattern φ_ζ for rapid recognition of test dynamical patterns.

5.2.3 Simulations

Consider the two dynamical patterns generated again from the Duffing oscillator [28,40]

$$\begin{aligned}\dot{x}_1 &= x_2 \\ \dot{x}_2 &= -p_2 x_1 - p_3 x_1^3 - p_1 x_2 + q \cos(wt)\end{aligned} \tag{5.4}$$

where $x = [x_1, x_2]^T$ is the state, p_1, p_2, p_3, w, and q are constant parameters, the system dynamics $f_2(x; p) = -p_2 x_1 - p_3 x_1^3 - p_1 x_2$ is an unknown, smooth nonlinear function, and $q \cos(wt)$ is a known periodic term which makes the behaviors of the Duffing oscillator more interesting [28].

The Duffing oscillator has been used in Chapter 4 as the reference model generating the recurrent reference trajectories. It is used here again because it can generate many types of dynamical behaviors, including periodic, quasiperiodic, and chaotic dynamical patterns. The periodic pattern and the chaotic pattern (shown in Figure 5.1, denoted as φ_ζ^1 and φ_ζ^2, respectively), are used to demonstrate the result of this section. Pattern φ_ζ^1 is generated from system (5.4), with initial condition $x(0) = [x_1(0), x_2(0)]^T = [0.0, -1.8]^T$ and system parameters $p_1 = 0.55$, $p_2 = -1.1$, $p_3 = 1.0$, $w = 1.8$, and $q = 1.498$. Pattern φ_ζ^2 is generated with the same system parameters except $p_1 = 0.35$.

The following dynamical RBF network, which is slightly modified from Equation (3.2), is employed to identify the unknown dynamics $f_2(x; p)$ of the two training dynamical patterns φ_ζ^1 and φ_ζ^2,

$$\dot{\hat{x}}_2 = -a_2(\hat{x}_2 - x_2) + \widehat{W}_2^T S(x) - q \cos(wt) \tag{5.5}$$

The RBF network $\widehat{W}_2^T S_2(x)$ is constructed in a regular lattice, with nodes $N = 441$, the centers μ_i evenly spaced on $[-3.0, 3.0] \times [-3.0, 3.0]$, and the widths $\eta_i = 0.3$. The weights of the RBF networks are updated according to Equation (3.5). The design parameters for Equations (5.5) and (3.5) are $a_2 = 5$, $\Gamma_2 = 2$, and $\sigma_2 = 0.001$. The initial weights $\widehat{W}_2(0) = 0.0$.

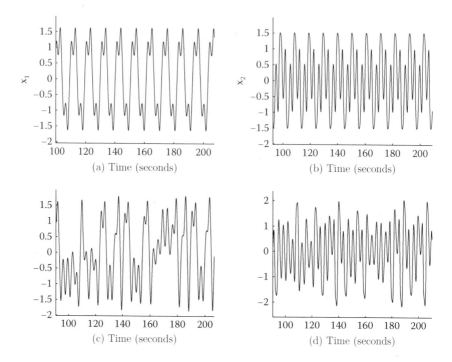

FIGURE 5.1
Periodic and chaotic dynamical patterns.

The phase portrait of dynamical pattern φ_ζ^1 is shown in Figure 5.2a. Its corresponding system dynamics $f_2(x; p)$ is shown in Figure 5.2b. Through deterministic learning, the system dynamics $f_2(x; p)$ of dynamical pattern φ_ζ^1 can be locally accurately identified. According to Theorem 3.1, exponential convergence of a closed-loop identification system, as well as the convergence of $\widetilde{W}_{\zeta2}$ (a subvector of $\widehat{W}_2$) is obtained. In Figure 5.2c, it is seen that some weight estimates (of the neurons whose centers are close to the orbit of the pattern) converge to constant values, whereas some other weight estimates (of neurons centered far away from the orbit) are almost zero. The locally accurate NN approximation of $f_2(x; p)$ along the orbit of the periodic pattern φ_ζ^1 is clearly shown in Figures 5.2d and e. In Figure 5.2f, dynamical pattern φ_ζ^1 is represented by the constant RBF network $\overline{W}_2 S(x)$. This representation is definitely time-invariant, based on the fundamental information of the system dynamics. It is also spatially distributed, involving a large number of neurons distributed along the orbit of the dynamical pattern. The NN approximation is accurate only in the vicinity of the periodic pattern. For the other region where the orbit of the pattern does not explore, no learning occurs, corresponding to the zero-plane in Figure 5.2f, that is, the small values of $\overline{W}_2^T S_2(x)$ in the unexplored area.

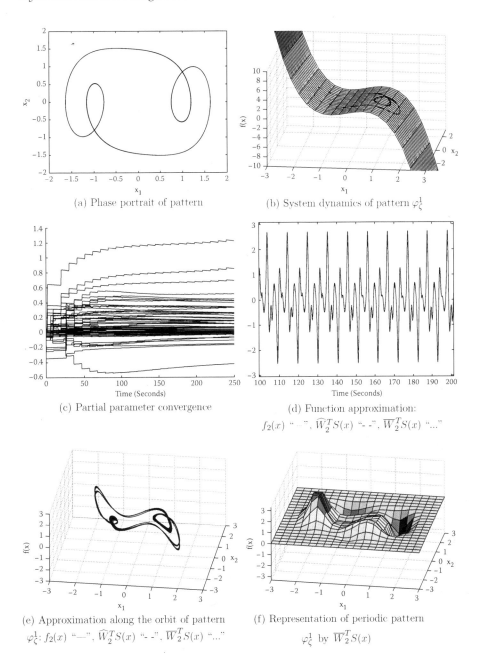

(a) Phase portrait of pattern

(b) System dynamics of pattern φ_ζ^1

(c) Partial parameter convergence

(d) Function approximation:
$f_2(x)$ "—", $\widehat{W}_2^T S(x)$ "- -", $\overline{W}_2^T S(x)$ "..."

(e) Approximation along the orbit of pattern
φ_ζ^1: $f_2(x)$ "—", $\widehat{W}_2^T S(x)$ "- -", $\overline{W}_2^T S(x)$ "..."

(f) Representation of periodic pattern
φ_ζ^1 by $\overline{W}_2^T S(x)$

FIGURE 5.2
Identification of periodic pattern φ_ζ^1.

Similarly, consider the chaotic pattern φ_ζ^2. Pattern φ_ζ^2 is generated from system (5.4), with initial condition $x(0) = [x_1(0), x_2(0)]^T = [0.3, -1.2]^T$ and system parameters $p_1 = 0.35$, $p_2 = -1.1$, $p_3 = 1.0$, $w = 1.8$, and $q = 1.498$. From Figures 5.3a and b, we can see the phase portrait and the system dynamics $f_2(x; p)$ of the chaotic pattern φ_ζ^2. The locally accurate NN approximation of system dynamics $f_2(x; p)$ along the orbit of the pattern is shown in Figures 5.3d and e. Figure 5.3f shows the time-invariant representation of chaotic pattern φ_ζ^2. It reveals that although the chaotic pattern φ_ζ^2 looks more complicated than the periodic pattern φ_ζ^1, the representation of a chaotic dynamical pattern can be processed in a similar way as that of a periodic dynamical pattern.

5.3   A Fundamental Similarity Measure

In temporal pattern recognition, the problem of characterizing the similarity between temporal or dynamical patterns is another important and difficult problem. In the literature of pattern recognition, there are many definitions for similarity of static patterns, most of which are based on distances, for example, Euclidean distance, Manhattan distance, and cosine distance [254]. To define the similarity of two dynamical patterns, the existing similarity measures developed for static patterns might become inappropriate. As dynamical patterns are defined as recurrent trajectories generated from nonlinear dynamical systems, it is known that small changes in initial states of the trajectory or system parameters may yield very different dynamical behaviors. This implies that it is rather difficult to characterize the similarity of two dynamical patterns via computing certain distances obtained simply from the time-varying states of the recurrent trajectories.

From the qualitative analysis of nonlinear dynamical systems [206,207], it is understood that the similarity between two dynamical behaviors lies in the *topological equivalence* and *structural stability* of two dynamical systems. Thus, the similarity of dynamical patterns is determined by the similarity of the system dynamics inherently within these dynamical patterns. In this chapter, we propose a similarity definition for dynamical patterns based on information from both system dynamics and pattern states: dynamical pattern A is similar to dynamical pattern B if (i) the state of pattern A stays within a local region of the state of pattern B, and (ii) the difference between the corresponding system dynamics along the state trajectory of pattern A is small. It is seen that the time dependence of dynamical patterns is excluded from the similarity definition.

To be specific, consider the dynamical pattern φ_ζ (as given by Equation [5.1]), and another dynamical pattern (denoted as $\varphi_\varsigma(x_{\varsigma 0}, p')$ or φ_ς) generated from the following nonlinear dynamical system:

$$\dot{x} = F'(x; p'), \qquad x(t_0) = x_{\varsigma 0} \tag{5.6}$$

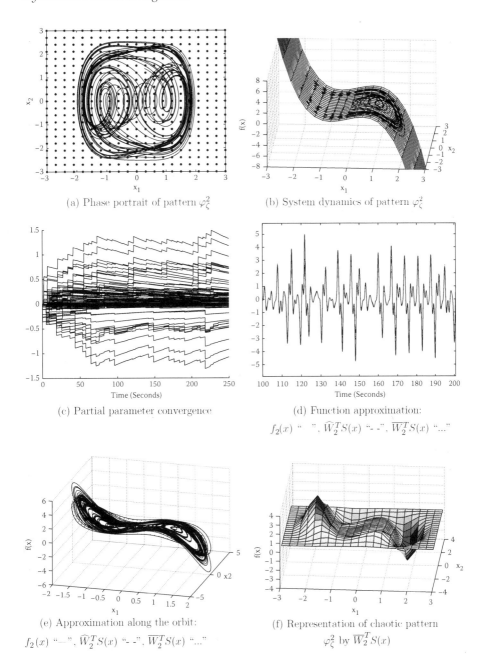

(a) Phase portrait of pattern φ_ζ^2

(b) System dynamics of pattern φ_ζ^2

(c) Partial parameter convergence

(d) Function approximation:
$f_2(x)$ "— ", $\widehat{W}_2^T S(x)$ "- -", $\overline{W}_2^T S(x)$ "..."

(e) Approximation along the orbit:
$f_2(x)$ "—", $\widehat{W}_2^T S(x)$ "- -", $\overline{W}_2^T S(x)$ "..."

(f) Representation of chaotic pattern
φ_ζ^2 by $\overline{W}_2^T S(x)$

FIGURE 5.3
Identification of chaotic pattern φ_ζ^2.

where the initial condition $x_{\varsigma 0}$, the system parameter vector p', and subsequently the nonlinear vector field $F'(x; p') = [f'_1(x; p'), \ldots, f'_n(x; p')]^T$, are possibly different from those for dynamical pattern φ_ζ. Because small changes in $x(t_0)$ or p' (or p in Equation [5.1]) may lead to large change of $x(t)$, it is clear that the similarity of dynamical patterns φ_ζ and φ_ς cannot be established by using only the time-varying states $x(t)$ of the patterns, or by some nonfundamental feature extracted from $x(t)$.

We propose the following definition of similarity for dynamical patterns.

DEFINITION 5.1

For two dynamical patterns φ_ς (given by Equation [5.6]) and φ_ζ (given by Equation [5.1]) consider the differences between the corresponding system dynamics along the orbit of pattern φ_ς, i.e., $\triangle f_i = |f_i(x; p) - f'_i(x; p')| \leq \varepsilon^*_i$ $(i = 1, \ldots, n)$, where ε^*_i is a finite positive constant. Dynamical pattern φ_ς is said to be similar to dynamical pattern φ_ζ if the state of pattern φ_ς stays within a neighborhood region of the state of pattern φ_ζ, and ε^*_i, called the similarity measure, is small.

In the above definition, no assumption is made about whether $f_i(x; p)$ or $f'_i(x; p')$ is available from measurement for characterizing the similarity. Suppose that through deterministic learning, system dynamics $f_i(x; p)$ $(i = 1, \ldots, n)$ of pattern φ_ζ has been accurately identified and effectively represented by constant RBF network $\overline{W}^T_i S_i(x)$ $(i = 1, \ldots, n)$. Based on the identification, we give the following definition characterizing how pattern φ_ς *is recognized* to be similar to pattern φ_ζ.

DEFINITION 5.2

For two dynamical patterns φ_ς (given by Equation [5.6]) and φ_ζ (given by Equation [5.1]) consider the approximate differences between the corresponding system dynamics along the orbit of pattern φ_ς; that is, $\triangle f_{Ni} = |\overline{W}^T_i S_i(x) - f'_i(x; p')| \leq \varepsilon^*_i + \xi^*_i$ $(i = 1, \ldots, n)$, where ε^*_i is a finite positive constant, and ξ^*_i is the approximation error given in Equation (5.2). Dynamical pattern φ_ς is recognized to be similar to dynamical pattern φ_ζ if the state of pattern φ_ς stays within a neighborhood region of the state of pattern φ_ζ, and $\varepsilon^*_i + \xi^*_i$, called the approximate similarity measure, is small.

Note that the differences $\triangle f_i$ and $\triangle f_{Ni}$ are given along the periodic or recurrent state of pattern φ_ς. Thus, they are functions of pattern state $x(t)$, and can be described simply using the L_∞ function norm

$$\|\triangle f_i\|_{t\infty} = \max_{x \in \varphi_\varsigma(x_{\varsigma 0}; p')} |f_i(x; p) - f'_i(x; p')|, \qquad i = 1, \ldots, n \qquad (5.7)$$

$$\|\triangle f_{Ni}\|_{t\infty} = \max_{x \in \varphi_\varsigma(x_{\varsigma 0}; p')} |\overline{W}^T_i S_i(x) - f'_i(x; p')|, \qquad i = 1, \ldots, n \qquad (5.8)$$

Another more appropriate description of $\triangle f_i$ and $\triangle f_{Ni}$ is to use the average L_p function norm:

$$\|\triangle f_i\|_{tp} = \left(\frac{1}{t}\int_{t_0}^{t_0+t} |f_i(x;p) - f_i'(x;p')|^p dt\right)^{1/p}, \qquad i = 1, \ldots, n \qquad (5.9)$$

$$\|\triangle f_{Ni}\|_{tp} = \left(\frac{1}{t}\int_{t_0}^{t_0+t} |\overline{W}_i^T S_i(x) - f_i'(x;p')|^p dt\right)^{1/p}, \qquad i = 1, \ldots, n \qquad (5.10)$$

where t_0 represents the initial time after a transient process. The most useful values of p are $p = 1, 2$.

REMARK 5.1
It is seen that the above similarity definitions are related to both the states and system dynamics of the two dynamical patterns. They are based on the time-invariant information of the system dynamics $f_i(x;p)$ and $f_i'(x;p')$ [or $\overline{W}_i^T S(x)$], which naturally include the information of system parameters. The state information (including initial states of dynamical patterns) is also involved. The above two definitions provide a reasonable way of measuring similarity between dynamical patterns.

REMARK 5.2
In contrast to the similarity definitions for static patterns, it is seen from Definitions 5.1 and 5.2 that pattern φ_ς being similar (or being recognized as similar) to pattern φ_ζ does not necessarily imply that the reverse is true. Moreover, in Definition 5.2, pattern φ_ς being recognized as similar to pattern φ_ζ refers to the case that correct recognition is based on accurate identification of pattern φ_ζ.

Note that in Definition 5.2, system dynamics $f_i'(x;p')$ of pattern φ_ς is still unavailable. We show in Section 5.4 that Definition 5.2 will be useful in providing an explicit measure of similarity in rapid recognition of pattern φ_ς.

5.4 Rapid Recognition of Dynamical Patterns

In this section, we investigate the mechanism for rapid recognition of dynamical patterns. To achieve recognition of a test pattern from a set of training patterns, one possible method is to identify the system dynamics and represent the test pattern by a constant RBF network (as done for training dynamical patterns through deterministic learning), and then compare the corresponding NN approximations with those of training dynamical patterns. One problem with such a method is that a direct comparison of NN approximations of system dynamics may be computationally demanding for the time available. For rapid recognition of a test dynamical pattern, it is preferred not to identify the system dynamics again, and complicated computations should be avoided as much as possible for easy and fast recognition.

Based on the time-invariant representation and the similarity measure, we propose a mechanism for rapid recognition of dynamical patterns. Using the constant RBF networks obtained in the identification phase, we construct a dynamical model for each training dynamical pattern. The constant RBF networks can quickly recall the learned knowledge by providing accurate approximations to the previously learned system dynamics of a training pattern. When a test pattern is presented to a dynamical model, a recognition error system is formed, which consists of the system generating the test pattern and the dynamical model corresponding to one of the training patterns. The recognition error system is in the simple form of a disturbed linear time-invariant (LTI) system, in which the differences of corresponding system dynamics are taken as bounded disturbances. Without identifying the system dynamics of the test pattern, and so without comparing system dynamics of corresponding dynamical patterns via numerical computation, a kind of *internal* and *dynamical* matching of system dynamics of the test and training pattern proceeds in the recognition error system. The state synchronization errors are proven to be approximately proportional to the differences of corresponding system dynamics. The test dynamical pattern is thus being recognized as similar to a training pattern if the state of the dynamical model synchronizes closely with the state of the test pattern. Thus, the synchronization errors can be taken as similarity measures between the test and the training dynamical patterns.

The recognition of a test dynamical pattern is achieved rapidly because the recognition process takes place from the beginning of measuring the state of the test pattern, without feature extraction from the test pattern (which is normally required in existing neural networks and statistical approaches for static pattern recognition [19,254]). The recognition process is automatically implemented with the evolution of the recognition error system. The significance of this approach is that the recognition process is a completely dynamical process with knowledge utilization. In other words, the problem of dynamical pattern recognition is turned into a problem of stability and convergence of a recognition error system.

5.4.1 Problem Formulation

Consider a training set containing dynamical patterns φ_ζ^k, $k = 1, \ldots, M$, with the kth training pattern φ_ζ^k generated from

$$\dot{x} = F^k(x, p^k), \qquad x(t_0) = x_{\zeta 0}^k \tag{5.11}$$

where p^k is the system parameter vector. As shown in Section 5.2, the system dynamics $F^k(x, p^k) = [f_1^k(x, p^k), \ldots, f_n^k(x, p^k)]^T$ can be accurately identified and stored in constant RBF networks $\overline{W}^{k^T} S_A(x) = [\overline{W}_1^{k^T} S_1(x), \ldots, \overline{W}_n^{k^T} S_n(x)]^T$.

Consider dynamical pattern φ_ς (as given by Equation [5.6]) as a test pattern. Without identifying the system dynamics of the test pattern φ_ς, the recognition problem is to search *rapidly* from the training dynamical patterns φ_ζ^k ($k = 1, \ldots, M$) for those *similar* to the given test pattern φ_ς in the sense of Definition 5.2.

5.4.2 Rapid Recognition via Synchronization

In the following, we present how rapid recognition of dynamical patterns is achieved. For the kth ($k = 1, \ldots, M$) training pattern φ_ζ^k, a dynamical model is constructed by using the time-invariant representation $\overline{W}^{k^T} S(x)$ as:

$$\dot{\bar{x}}^k = -B(\bar{x}^k - x) + \overline{W}^{k^T} S_A(x) \qquad (5.12)$$

where $\bar{x}^k = [\bar{x}_1^k, \ldots, \bar{x}_n^k]^T$ is the state of the dynamical (template) model, x is the state of an input test pattern φ_ς generated from Equation (5.6), and $B = diag\{b_1, \ldots, b_n\}$ is a diagonal matrix that is kept the same for all training patterns. Note that b_i ($1 \le i \le n$) is not chosen as a large value. Then, corresponding to the test pattern φ_ς and the dynamical model (5.12) (for training pattern φ_ζ^k), we obtain the following recognition error system:

$$\dot{\tilde{x}}_i^k = -b_i \tilde{x}_i^k + \overline{W}_i^{k^T} S_i(x) - f_i'(x, p'), \qquad i = 1, \ldots, n \qquad (5.13)$$

where $\tilde{x}_i^k = \bar{x}_i^k - x_i$ is the state tracking (or synchronization) error. It is clear that system (5.13) is in the simple form of a linear time-invariant system with bounded disturbance.

Note that without identifying the system dynamics of the test pattern φ_ς, the difference on system dynamics of the test and training patterns, that is, $|\overline{W}_i^{k^T} S_i(x) - f_i'(x, p')|$, is not available from direct computation. Nevertheless, it will be shown that the difference between system dynamics can be explicitly measured by $|\tilde{x}_i^k|$. Thus, if the state $\bar{x}_i^k$ of the dynamical model (5.12) tracks closely to (or synchronizes with) the state x of dynamical pattern φ_ς, that is, $|\tilde{x}_i^k|$ is small, then the test pattern φ_ς can be recognized as similar to the training pattern φ_ζ^k in the sense of Definition 5.2.

THEOREM 5.1
Consider the recognition error system (5.13) corresponding to test pattern φ_ς and the dynamical model (5.12) for training pattern φ_ζ^k. Then, the synchronization errors $\tilde{x}_i^k$ ($i = 1, \ldots, n$) converge exponentially to a neighborhood of zero. Furthermore, for finite T, $|\tilde{x}_i^k|_{t \ge T}$ is approximately proportional to the difference between the system dynamics of test pattern φ_ς and the identified system dynamics of training pattern φ_ζ^k.

PROOF To simplify the notion, we remove the superscript $(\cdot)^k$ in the following derivations.

For the recognition error system (5.13), consider Lyapunov function $V_i = \frac{1}{2}\tilde{x}_i^2$. Its derivative is

$$\dot{V}_i = \tilde{x}_i \dot{\tilde{x}}_i = -b_i \tilde{x}_i^2 - \tilde{x}_i \left(\overline{W}_i^T S_i(x) - f_i'(x; p')\right)$$

Note

$$-\frac{1}{2}b_i \tilde{x}_i^2 - \tilde{x}_i \left(\overline{W}_i^T S_i(x) - f_i'(x; p')\right)$$

$$\leq -\frac{1}{2}b_i \tilde{x}_i^2 + |\tilde{x}_i| \left|\overline{W}_i^T S_i(x) - f_i'(x; p')\right|$$

$$\leq \frac{\left|\overline{W}_i^T S_i(x) - f_i'(x; p')\right|^2}{2b_i} \tag{5.14}$$

Then, we have

$$\dot{V}_i \leq -\frac{1}{2}b_i \tilde{x}_i^2 + \frac{\left(\overline{W}_i^T S_i(x) - f_i'(x; p')\right)^2}{2b_i}$$

$$= -b_i V_i + \frac{\Delta f_{Ni}^2}{2b_i} \tag{5.15}$$

Denote $\rho_i := \frac{\|\Delta f_{Ni}\|_\infty^2}{2b_i^2}$. Then, Equation (5.15) gives

$$0 \leq V_i(t) < \rho_i + (V_i(0) - \rho_i)\exp(-b_i t) \tag{5.16}$$

From (5.16), we have

$$\tilde{x}_i^2 < 2\rho_i + 2V_i(0)\exp(-b_i t) \tag{5.17}$$

which implies that given $v_i > \sqrt{2\rho_i}$, there exists a finite time T, such that for all $t \geq T$, the state tracking error $\tilde{x}_i(t)$ will converge exponentially to a neighborhood of zero, that is, $|\tilde{x}_i|_{t \geq T} \leq v_i$, with the size of the neighborhood v_i approximately proportional to $\frac{\|\Delta f_{Ni}\|_\infty}{b_i}$, that is, approximately proportional to $\xi_i^* + \varepsilon_i^*$, and inversely proportional to b_i. Thus, we have that $|\tilde{x}_i|_{t \geq T}$ (for finite T) is approximately proportional to the difference between the system dynamics $f_i'(x, p')$ of test pattern φ_ς and the identified system dynamics $\overline{W}_i^{k^T} S_i(x)$ of training pattern φ_ς^k. ∎

We noted that the difference between system dynamics of the test and training patterns is not available from direct computation. From the above analysis, it is seen that the difference between the system dynamics of the test and training patterns can be explicitly measured by $|\tilde{x}_i|_{t \geq T}$. Thus, we take the following method to rapidly recognize a test dynamical pattern from a set of training dynamical patterns:

1. Identify the system dynamics of a set of training dynamical patterns $\varphi_\zeta^k \ k = 1, \ldots, M$.

2. Construct a set of dynamical models (5.12) for the training dynamical patterns φ_ζ^k.

3. Take the state $x(t)$ of a test pattern φ_ζ as the RBFN input to the dynamical models (5.12), and compute the average L_p norm of the state estimation error $\tilde{x}_i^k(t)$, for example, for $p = 1$,

$$\|\tilde{x}_i^k(t)\|_{t1} = \frac{1}{t} \int_{t_0}^{t_0+t} |\tilde{x}_i^k(t)| dt, \qquad i = 1, \ldots, n \qquad (5.18)$$

4. Take the training dynamical pattern whose corresponding dynamical model yields the smallest $\|\tilde{x}_i^k\|_{t1}$ as the one most *similar* to the test dynamical pattern φ_ζ in the sense of Definition 5.2.

REMARK 5.3

It is seen that the recognition is achieved due to the internal matching of system dynamics according to $|\overline{W}_i^{k^T} S_i(x) - f_i(x, p')|$, by utilizing the time-invariant and spatially distributed representation and the similarity definition, which contain complete information on both states and system dynamics of dynamical patterns. Recognition of a dynamical pattern is converted into a problem of stability and convergence of a disturbed linear time-invariant recognition error system (5.13). The recognition is automatically implemented with the convergence of the recognition error system (5.13), and the outcome of the process; that is, the synchronization error $|\tilde{x}_i|$, is naturally taken as the measure of similarity between the test and training patterns. The representation, the similarity definition, and the recognition mechanism are three important elements to the proposed recognition approach for dynamical patterns.

REMARK 5.4

The recognition of a test pattern φ_ζ from a set of training patterns φ_ζ^k ($k = 1, \ldots, M$) is achieved in a parallel, rapid, and dynamic manner: (i) a recognition system is built up by using a set of dynamical models, each of them representing one training dynamical pattern, and recognition of the test pattern φ_ζ from the set of training patterns φ_ζ^k will proceed in a parallel way. (ii) Recognition of the test pattern φ_ζ occurs rapidly, because the recognition process takes place from the beginning of measuring the state x of the test pattern, and ends within one period T of the recurrent trajectory of the test pattern; moreover, because the recognition proceeds in a parallel manner, the time of recognizing the test pattern from a large number of training patterns will be the same as from a few (e.g., two) training patterns. (iii) The recognition process does not need any feature extraction procedure for the test dynamical pattern. It also does not need to compare the states or system dynamics of the test pattern with those of the set of training patterns by any form of static numerical computation. Recognition of a dynamical pattern is achieved in a completely dynamic manner.

5.4.3   Simulations

To verify the rapid recognition approach, we take dynamical patterns φ_ζ^1 and φ_ζ^2 used in Section 5.2 as two training dynamical patterns. Using the time-invariant representations $\overline{W}_2^{k^T} S_2(x)$ $(k = 1, 2)$ obtained in Section 5.2, two dynamical models are constructed according to (5.12) for the two training patterns as

$$\dot{\bar{x}}_2^k = -b_2(\bar{x}_2^k - x_2) + \overline{W}_2^{k^T} S_2(x) - q\cos(wt); \qquad k = 1, 2 \qquad (5.19)$$

where $\bar{x}_2^k$ is the state of the dynamical model for training pattern φ_ζ^k, x_2 is the state of the test pattern described below, and $b_2 > 0$ is a design constant, which should not be a large value ($b_2 = 2$ in this section).

Two periodic patterns and one chaotic pattern, as shown in Figure 5.4, are used as the test patterns φ_ζ^1, φ_ζ^2; and φ_ζ^3. Test pattern φ_ζ^1 is generated from system (5.4), with initial condition $x(0) = [x_1(0), x_2(0)]^T = [0.0, -1.8]^T$ and system parameters $p_1 = 0.6$, $p_2 = -1.1$, $p_3 = 1.0$, $w = 1.8$, and $q = 1.498$. Test patterns φ_ζ^2 and φ_ζ^3 are also generated from system (5.4). The initial condition and system parameters of test patterns φ_ζ^2 and φ_ζ^3 are the same as those of test pattern 1, except that $p_1 = 0.4$ and $p_1 = 0.33$, respectively.

First, consider the recognition of test pattern φ_ζ^1 by training patterns φ_ζ^1 and φ_ζ^2. Figures 5.5a and b show the system dynamics $f_2(x; p) = -p_2 x_1 - p_3 x_1^3 - p_1 x_2$ along the orbit of test pattern φ_ζ^1, together with the RBFN approximations of the system dynamics of the training patterns φ_ζ^1 and φ_ζ^2, respectively. The state synchronization or estimation errors $\tilde{x}_2^k(t)$ $(k = 1, 2)$, are shown in Figure 5.5c and d. The average l_1 norms of the synchronization errors, that is, $\|\tilde{x}_2^k(t)\|_{t1}$ $(k = 1, 2)$, are shown in Figures 5.5e and f. It is clearly seen in Figure 5.5f that from the beginning stage of the recognition process, $\|\tilde{x}_2^1(t)\|_{t1}$ is smaller than $\|\tilde{x}_2^2(t)\|_{t1}$. Thus, the test pattern φ_ζ^1 is rapidly recognized as more similar to training pattern φ_ζ^1 than to training pattern φ_ζ^2.

Similarly, in recognition of test dynamical pattern φ_ζ^2, it is seen from Figure 5.6 that the test pattern φ_ζ^2 is more similar to the chaotic training pattern φ_ζ^2 than to the periodic training pattern φ_ζ^1. It is also seen from Figure 5.7 that the test chaotic pattern φ_ζ^3 is more similar to the chaotic training pattern φ_ζ^2 than to the periodic training pattern φ_ζ^1. From Figures 5.5 to 5.7, we can see that the recognition of the test dynamical patterns occurs quickly within a very short period of time. Figures 5.5 through 5.7 also reveal that the chaotic training pattern φ_ζ^2 is more representative than the periodic training pattern φ_ζ^1 in rapid recognition of test dynamical patterns.

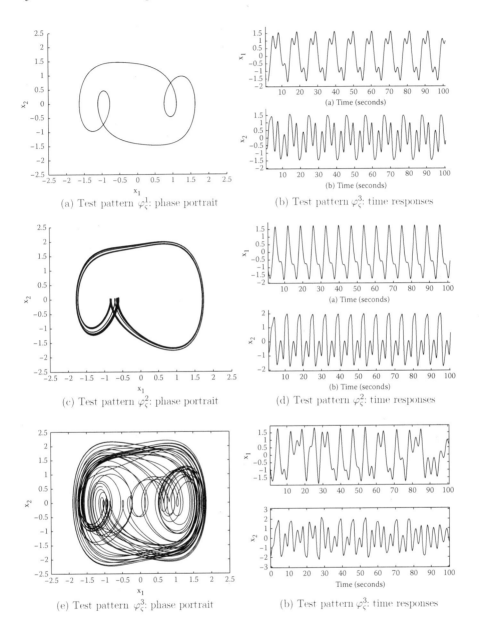

(a) Test pattern φ_ς^1: phase portrait

(b) Test pattern φ_ς^3: time responses

(c) Test pattern φ_ς^2: phase portrait

(d) Test pattern φ_ς^2: time responses

(e) Test pattern φ_ς^3: phase portrait

(b) Test pattern φ_ς^3: time responses

FIGURE 5.4
Test dynamical patterns.

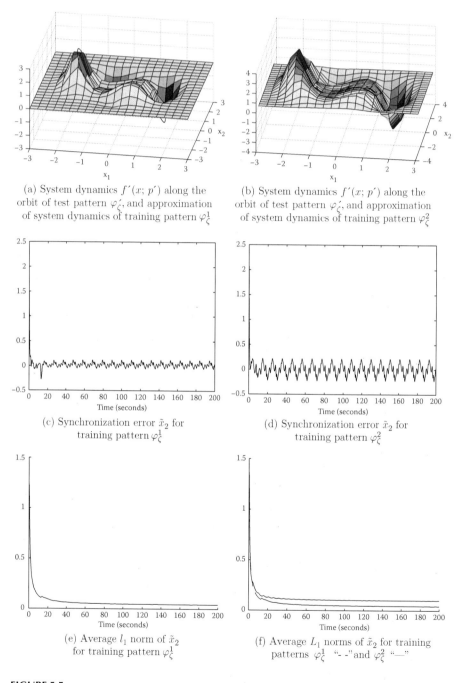

(a) System dynamics $f'(x; p')$ along the orbit of test pattern φ_ζ', and approximation of system dynamics of training pattern φ_ζ^1

(b) System dynamics $f'(x; p')$ along the orbit of test pattern φ_ζ', and approximation of system dynamics of training pattern φ_ζ^2

(c) Synchronization error $\tilde{x}_2$ for training pattern φ_ζ^1

(d) Synchronization error $\tilde{x}_2$ for training pattern φ_ζ^2

(e) Average l_1 norm of $\tilde{x}_2$ for training pattern φ_ζ^1

(f) Average L_1 norms of $\tilde{x}_2$ for training patterns φ_ζ^1 "- -" and φ_ζ^2 "—"

FIGURE 5.5
Recognition of test pattern φ_ζ^1 by training patterns φ_ζ^1 and φ_ζ^2.

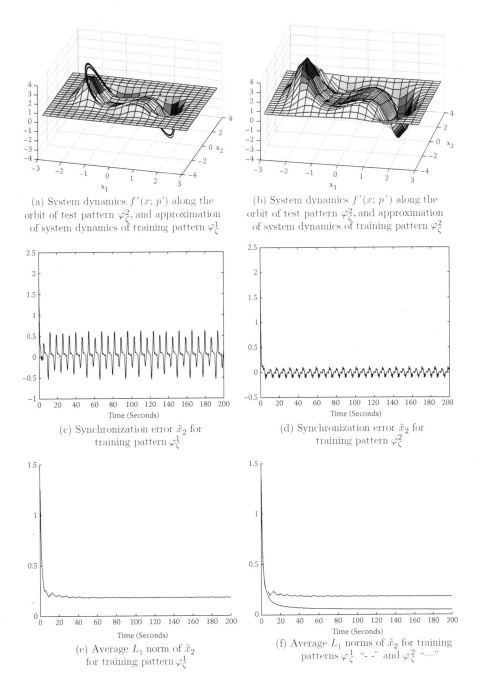

(a) System dynamics $f'(x; p')$ along the orbit of test pattern φ_ς^2, and approximation of system dynamics of training pattern φ_ς^1

(b) System dynamics $f'(x; p')$ along the orbit of test pattern φ_ς^2, and approximation of system dynamics of training pattern φ_ς^2

(c) Synchronization error $\tilde{x}_2$ for training pattern φ_ς^1

(d) Synchronization error $\tilde{x}_2$ for training pattern φ_ς^2

(e) Average L_1 norm of $\tilde{x}_2$ for training pattern φ_ς^1

(f) Average L_1 norms of $\tilde{x}_2$ for training patterns φ_ς^1 "- -" and φ_ς^2 "—"

FIGURE 5.6
Recognition of test pattern φ_ς^2 by training patterns φ_ς^1 and φ_ς^2.

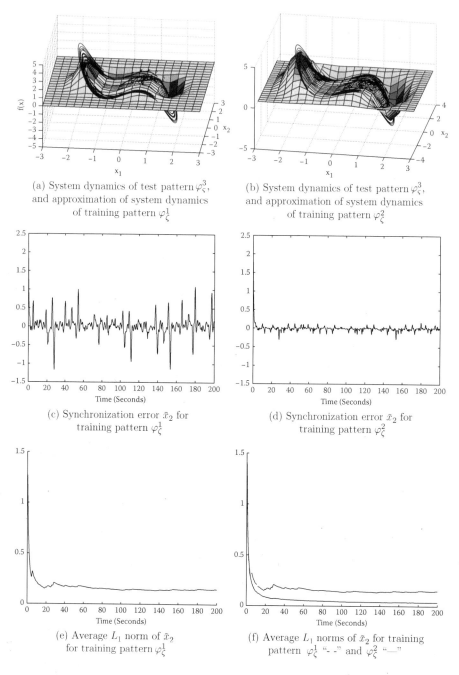

(a) System dynamics of test pattern φ_ζ^3, and approximation of system dynamics of training pattern φ_ζ^1

(b) System dynamics of test pattern φ_ζ^3, and approximation of system dynamics of training pattern φ_ζ^2

(c) Synchronization error $\tilde{x}_2$ for training pattern φ_ζ^1

(d) Synchronization error $\tilde{x}_2$ for training pattern φ_ζ^1

(e) Average L_1 norm of $\tilde{x}_2$ for training pattern φ_ζ^1

(f) Average L_1 norms of $\tilde{x}_2$ for training pattern φ_ζ^1 "- -" and φ_ζ^2 "—"

FIGURE 5.7
Recognition of test pattern φ_ζ^3 by training patterns φ_ζ^1 and φ_ζ^2.

5.5 Dynamical Pattern Classification

With the results on identification, representation, and recognition of dynamical patterns, in this section we further investigate the construction of the recognition systems for classification [239]. The problem is to assign a test dynamical pattern, φ_ς, to one of N classes $\Psi_1, \ldots, \Psi_N$, based on the predefined similarity measure.

The recognition system is constructed as consisting of many dynamical (template) models, as described in Equation (5.12). Each of the dynamical models is a dynamical RBF network representing one training dynamical pattern. Moreover, each class of dynamical patterns is represented by a set of chosen dynamical pattern templates or (prototypes), and is described by the corresponding template dynamical models. As the similarity between two dynamical patterns lies in the *topological similarity* of their underlying system dynamics, and the similarity distances between various dynamical patterns can be accurately measured by using the synchronization errors, the recognition system can be built up according to how the template dynamical models are arranged, that is, in a specific order according to the qualitative analysis of nonlinear dynamical systems [206] and the principle of minimal distance or nearest-neighbor classification [95]. We show that a hierarchical structured knowledge representation is set up based on the similarity of system dynamics, in which the concepts of topological equivalence, structural stability, bifurcation, and chaos all together provide an inclusive classification of various types of dynamic patterns.

The recognition system presented in this section can not only classify into different classes of dynamical patterns but can also distinguish a set of dynamical patterns generated from the same class. It can also be designed to identify bifurcation points, which actually form the boundaries between different subclasses of a set of dynamical patterns. The result of this chapter provides mathematical insight into some recent hypotheses on the roles of synchronization and chaos in brain science [54,210]. It also shows that the mechanism for the human recognition process, although not fully understood, is seemingly consistent with the mechanisms for deterministic learning and dynamical pattern recognition studied here.

5.5.1 Nearest-Neighbor Decision

The nearest-neighbor decision rule is a commonly used classification algorithm in pattern recognition [95], in which each class is represented by a set of chosen templates (or prototypes). When an unknown pattern is to be classified, its closest neighbor (with minimum distance) is found from among all the templates, and the class label is decided accordingly. If the number of preclassified prototypes is large, it makes good sense to use, instead of the single nearest neighbor, the majority vote of the nearest k neighbors. This method is referred to as the k-nearest-neighbor rule [95]. (The value of k should be odd to avoid ties on class-overlap regions.)

For dynamical pattern recognition, we propose that the nearest-neighbor classification, among many existing classification algorithms [95], is particularly suitable due to the following reasons:

1. Through deterministic learning, each training dynamical pattern can be accurately identified and correctly classified into categories. This situation is in accordance with the principle of nearest-neighbor classification, which normally does not assume any statistical knowledge of the distribution of the test pattern and categories, and depends only on a collection of correctly classified training samples [95].

2. According to the recognition mechanism, the similarity distance between the test and training dynamical patterns can be measured by their synchronization errors, for example, $\|\tilde{x}_i\|_{t1}$. Note that by deterministic learning, the fundamental information extracted from a dynamical pattern is described in a spatially distributed manner. Therefore, it is difficult to represent a dynamical pattern by a feature vector in a d-dimensional pattern space as in traditional pattern recognition. Nevertheless, the availability of the similarity distance makes it natural to use the minimal distance or nearest-neighbor decision rule for dynamical pattern classification.

3. The main problem with using the nearest-neighbor classification is the computational complexity caused by the large number of distance computations, in which all the distances between the input pattern and the prototype patterns are computed. For realistic pattern space dimensions, it is hard to find any variation of the rule that would be significantly lighter than the brute force method [95]. This major problem can be easily solved when the dynamical models within the recognition system are disposed in a parallel structure, such that by using the recognition mechanism described in Section 5.4, the similarity distances between the test pattern and all the training patterns are generated automatically and simultaneously in a dynamical recognition process.

5.5.2 Qualitative Analysis of Dynamical Patterns

As mentioned above, the recognition system is to be constructed with the dynamical models being arranged in some specific order. This specific order can be designed according to the qualitative analysis of nonlinear dynamical systems [120,206,207], in which the concepts of topological equivalence of dynamical systems, structural stability, bifurcation, and chaos together provide an inclusive classification of various types of dynamic patterns. In particular:

1. The concept of topological equivalence of dynamical systems is proposed for the purpose of studying qualitative features of the

behavior of different dynamical systems. Two dynamical systems are considered as topologically equivalent if their phase portraits are qualitatively similar, namely, if one portrait can be obtained from another by a continuous transformation [120,206]. The concept of topological equivalence can be used to define structural stability, which describes dynamical behaviors whose phase portraits do not change qualitatively under sufficiently small perturbations on system dynamics. More specifically, for a dynamical system to be structurally stable, it means that any system with sufficiently close system dynamics is topologically equivalent to the given one [207]. Thus, recurrent trajectories generated from structurally stable systems are similar in the sense of Definition 5.1, and it is reasonable to say that structurally stable dynamical patterns belong to the same subclass.

2. Whereas topological equivalence is related to structural stability, the concept of topological nonequivalence yields bifurcation. When the parameters of a dynamical system change, the appearance of a topologically nonequivalent phase portrait is called a bifurcation. Thus, a bifurcation is a change of the topological type of dynamical behaviors as a parameter-dependent dynamical system varies its parameters across a critical value referred to as a bifurcation point [120]. Bifurcation points form the bifurcation boundaries where structural instability occurs. From our point of view, the bifurcation boundaries can be taken as the boundaries between different subclasses of dynamical patterns.

3. A bifurcation diagram is a stratification of its parameter space induced by the topological equivalence, together with representative phase portraits for each stratum. A bifurcation diagram classifies in a very condensed way all possible modes of behavior of dynamical systems and transitions between them under parameter variations. The bifurcation diagram of even a simple dynamical system may be very complicated, composing an infinite number of strata. Nonetheless, only partial knowledge of the bifurcation diagram still provides essential information on the dynamical behaviors of the dynamical system [120]. Therefore, the bifurcation diagram can be naturally taken as a classification diagram for dynamical behaviors and for dynamical patterns.

Thus, all these elegant concepts from qualitative analysis of dynamical systems can be useful to arrange the dynamical models into a specific order in the recognition system construction.

5.5.3 A Hierarchical Structure

Assuming that the nearest-neighbor decision rule is used, each class is represented by a set of chosen templates (or prototypes). To save memory space, it is desirable not to store all the identified training patterns as templates.

Subsequently, an important question is how to choose the most representative patterns as appropriate templates, such that the number of templates may be decreased without losing accuracy.

As stated in Chapter 3, all the dynamical patterns undergoing recurrent motions, including quasi-periodic and chaotic ones, can be accurately identified by deterministic learning. Compared with the periodic patterns, quasi-periodic and chaotic patterns are more spatially expanded, and usually occur under a slight parameter variation. This means that the dynamical models corresponding to quasi-periodic and chaotic patterns are very suitable for use as template models in the recognition system. Specifically, at the first level of the hierarchical structure, a few chaotic patterns are chosen as templates, according to the bifurcation diagrams, to represent classes of dynamical patterns in a broad sense. In the subsequent levels, quasi-periodic and periodic patterns are used to represent classes and subclasses of dynamical patterns. In this way, the recognition system is constructed with the dynamical template models being arranged according to a hierarchical structured knowledge representation based on the similarity of system dynamics.

For demonstration, it is seen from Figures 5.6 and 5.7 that the test patterns 2 and 3 are being recognized as more similar to the chaotic training pattern 2, rather than similar to the periodic training pattern 1. From Figure 5.5, it is seen that the test pattern 1 is being recognized as more similar to the periodic training pattern 1, and also similar to the chaotic training pattern 2 (because the difference is small). Thus, it is revealed that the chaotic training pattern 2 is more representative than the periodic one, and the corresponding (chaotic) dynamical model can be taken as the template model and be arranged in the first level in the recognition system construction. On the other hand, the periodic and quasi-periodic training patterns are also useful, because the corresponding dynamical models can be used in the subsequent levels to improve the classification accuracy and the discrimination capability.

REMARK 5.5

The results may provide support to the "dynamical hypothesis" in cognitive science [228]: Natural cognitive systems are certain kinds of dynamical systems, and are best understood from the perspective of dynamics. In the proposed approach, it has been shown that identification, recognition, and classification of dynamical patterns are indeed best understood from a viewpoint of stability analysis of linear time-varying or linear time-invariant systems, using concepts and theories from system identification, adaptive control, and dynamical systems. The result may also provide mathematical insight into the hypotheses on the roles of synchronization and chaos in brain science. For example, it is stated in [54] that "The brain transforms sensory messages into conscious perceptions almost instantly. Chaotic collective activity involving millions of neurons seems essential for such rapid recognition." These hypotheses can be reasonably interpreted by our results on the representation, recognition, and classification of dynamical patterns.

REMARK 5.6

It is clear that the implementation of a comprehensive recognition system requires the ability to integrate a large number of dynamical models, and very large-scale circuits in silicon. This is becoming less of an issue than previously with the rapid development of microelectronics, especially on VLSI technologies.

5.6 Summary

In this chapter, we have proposed an approach for rapid recognition of dynamical patterns. The elements of the recognition approach include: (i) a time-invariant and spatially distributed representation for dynamical patterns; (ii) a similarity measure based on system dynamics; and (iii) a mechanism in which rapid recognition of dynamical patterns is achieved by state synchronization. It has been shown that a time-varying dynamical pattern can be effectively represented by using complete information on its state trajectory and its underlying system dynamics along the state trajectory. Based on the proposed similarity measure for dynamical patterns, a mechanism for rapid recognition of dynamical patterns has been presented. Rapid recognition can be automatically implemented in a dynamical recognition process without conventional feature extraction. The outcome of the recognition process, that is, the synchronization error, is naturally taken as the measure of similarity between the test and training patterns. The dynamical recognition process does not need to compare directly the states or system dynamics of the test and training patterns by any form of numerical computation.

The proposed recognition approach can facilitate construction of recognition systems for dynamical pattern classification. The constructed recognition system promises to be able to classify different classes of dynamical patterns, and distinguish a set of dynamical patterns generated from the same class. It can also be designed to detect bifurcation, which is an important task for many industrial applications. Moreover, the proposed approach appears to be consistent with mechanisms of human recognition of temporal patterns, and may provide insight to natural cognitive systems from the perspective of dynamics. It presents a new model for information processing, that is, dynamical parallel distributed processing (DPDP). When implemented in a hybrid analog–digital manner, DPDP will increase significantly the computational efficiency for information processing in uncertain dynamic environments.

6

Pattern-Based Intelligent Control

6.1 Introduction

Pattern recognition was studied in the control literature in the 1960s together with adaptive, learning, and self-organizing systems; see, for instance, [226]. In that time, a pattern in control was defined as a control situation that was represented by a set of state variables. Information on a control situation learned during the process of closed-loop control was taken as a control experience. Pattern recognition techniques were proposed to classify different control situations. Based on the classification result, an experienced controller corresponding to the specific control situation was selected to control the system [56].

The idea of using pattern recognition to achieve an advanced intelligent control might be motivated naturally by human learning and control, in which pattern identification, recognition, and control together play important roles. It has been observed that with sufficient practice a human can learn many highly complicated control tasks, and these tasks can be performed again and again by a proficient individual with little effort. The implementation of the idea in technology, however, is very difficult. One problem, which was indicated as early as in 1970 by Fu [56], is learning in nonstationary or dynamic environments. This might be the most difficult problem in the area of adaptive and learning control systems. Other problems include representation, rapid recognition, and classification of different patterns in control, that is, control situations. It is obvious that conventional pattern recognition methods, for example, representation of nonstationary state variables by using a finite number of different stationary patterns, and recognition techniques for identification and classification of stationary patterns, are not suitable to cope with these problems. A new framework is required to implement pattern identification, recognition, and control in a unified way.

The deterministic learning (DL) theory presented in Chapters 3 to 5 provides elements toward a new framework for pattern-based learning control. Through deterministic learning, the system dynamics of nonlinear dynamical systems can be locally accurately identified. An appropriately designed adaptive NN controller is shown capable of learning the closed-loop system dynamics during tracking control to recurrent reference trajectory. The learned knowledge is represented as a time-invariant NN approximation and is stored in a constant RBF network. Moreover, a DL-based approach is proposed for representation, similarity definition, and rapid recognition of dynamical patterns. It is shown that dynamical patterns can be effectively represented and stored in a time-invariant manner using a locally accurate NN approximation of system dynamics. A similarity definition for dynamical patterns is also given based on system dynamics. Based on the time-invariant representation and the similarity definition, a scheme is proposed in which rapid recognition of dynamical patterns can be implemented via state estimation.

In this chapter, based on the aforementioned results, we propose a framework for pattern-based intelligent control as follows. First, for different training control tasks, the system dynamics corresponding to the training control tasks are identified via deterministic learning. A set of training dynamical patterns is defined based on the identification. The representation and similarity of dynamical patterns are also presented. A set of pattern-based NN controllers is constructed accordingly. Second, a dynamical pattern classification system is introduced that can rapidly recognize dynamical patterns and switch quickly among the set of pattern-based NN controllers. For a test control task, if the corresponding dynamical pattern is recognized as very similar to one previous training pattern, then the NN controller corresponding to the training pattern is selected and activated. Third, the selected NN learning controller is used which can effectively exploit the learned knowledge to achieve improved control performance without readapting to the uncertainties in the closed-loop control process. This can be regarded as the advantage of knowledge utilization in dynamical environments. Note that if the control task corresponds to a dynamical pattern not experienced before, the identification process (as in the first step) will be restarted. Time permitting, the learned knowledge will yield a new NN controller which will be added to the set of pattern-based NN controllers.

This chapter extends some earlier work by the authors in [243,247].

6.2   Pattern-Based Control

6.2.1   Definitions and Problem Formulation

Consider the system model

$$\begin{cases} \dot{x}_1 = x_2 \\ \dot{x}_2 = f^k(x) + u \end{cases} \tag{6.1}$$

where $x = [x_1, x_2]^T \in R^2$, $u \in R$ are the state variables and system input, respectively, $f^k(x)$ $(k = 0, 1, \ldots, K)$ are the unknown smooth nonlinearities, corresponding to different operating environments such as a normal state $(k = 0)$ and changes in system dynamics (or system parameters), faults in the system, sensor failures, and external disturbances $(k = 1, \ldots, K)$. The control task is tracking control of the system state $x(t)$ in all the environments to a set of periodic or periodic-like reference orbits $x_d(t)$ generated from the following reference models:

$$\begin{cases} \dot{x}_{d_1} = x_{d_2} \\ \dot{x}_{d_2} = f_d^m(x_d) \end{cases} \tag{6.2}$$

where $x_d = [x_{d_1}, x_{d_2}]^T \in R^2$ is the system state and $f_d^m(\cdot)$ $(m = 1, \ldots, M)$ is a smooth nonlinear function. There are different reference tracking orbits x_d^m corresponding to changes in initial conditions or system parameters.

Obviously, two types of dynamical patterns exist in the tracking control process. They are referred to as *reference dynamical patterns* and *closed-loop dynamical patterns*. The definitions of the two types of dynamical patterns are as follows.

DEFINITION 6.1
A *reference dynamical pattern* is defined as a recurrent reference system trajectory $x_d(t)(\forall t \geq 0)$ generated from the reference model. It is started from initial condition $x_d(0)$ and is denoted as φ_d for concise presentation.

DEFINITION 6.2
A *closed-loop dynamical pattern* is defined as a recurrent system state trajectory $x(t)$ generated from closed-loop tracking control to a recurrent reference trajectory. It is started from initial condition x_0, and is denoted as φ_ζ.

REMARK 6.1
The reference dynamical pattern is related to the control task, but not related to the plant and the controller. The closed-loop dynamical pattern is related to the control task, that is, tracking to a recurrent reference orbit, the corresponding controller, and the closed-loop system dynamics.

The pattern-based control structure consists of a phase of identification and another phase of recognition and control. More specifically, the objective of pattern-based control is twofold: (i) to identify the system dynamics of dynamical patterns as well as the corresponding control dynamics, and construct a set of pattern-based NN controllers by using the obtained control system dynamics; and (ii) to rapidly recognize and classify dynamical patterns, and select a pattern-based NN controller based on the classification to achieve guaranteed stability and performance.

6.2.2   Control Based on Reference Dynamical Patterns

Assume that there exist m reference dynamical patterns generated from reference model (6.2) in the control process, and the system dynamics $f^k(x)$ of the plant (6.1) remains unchanged; that is, $f^k(x) \equiv f(x)$ for all k.

In this case, the pattern-based control process consists of the following steps:

1. Identify the local system dynamics $f_d^m(x_d)$ of the reference dynamical patterns. This can be conducted in the same way as in Chapter 3. The identified reference patterns are represented as shown in Chapter 5 by the locally accurate NN approximation $f_{d_{nn}}^m(x_d)$ achieved in a local region along the recurrent orbit $x_d^m(t)$.

2. Identify the local controlled system dynamics $f(x)$ corresponding to each reference dynamical pattern. This can be conducted in the same way as in Chapter 4. The identified results are represented by the locally accurate NN approximation $f_{nn}^m(x)$ achieved in a local region along the recurrent orbit $x(t)$ when $x(t) \to x_d(t)$. A set of pattern-based NN controllers is constructed accordingly by using the obtained control system dynamics as follows (see Equation [4.3]);

$$u^m = -z_1 - c_2 z_2 - f_{nn}^m(x) + \dot{\alpha}_1 \tag{6.3}$$

where $m = 1, \ldots, M$.

3. Construct dynamic models using $f_{d_{nn}}^m(x_d)$ and rapidly recognize a test reference dynamical pattern via state synchronization or estimation. This can be conducted in the same way as in Chapter 5. The estimator with the smallest estimation error corresponds to the training reference dynamical pattern which is most similar to the test reference dynamical pattern.

4. Select the corresponding NN controller based on rapid recognition and classification. The selected NN controller will be able to achieve guaranteed stability and improved control performance.

Different tracking control tasks for the same system, that is, different reference orbits, will give many control system dynamics $f_{nn}^m(x)$ as local models. These local models can be merged to form a unified model $f_{nn}(x)$ valid for a larger region, which implies that past experiences can be combined to make up an "overall" experience. The overall experience clearly demonstrates the "learning from experience" paradigm in AI [257]: the more experiences we derive from some specific region, the better we would learn the system in that region.

This result can be extended to pattern-based control of more general nonlinear systems as studied in Chapter 4.

6.2.3 Control Based on Closed-Loop Dynamical Patterns

Assume that there exist k closed-loop dynamical patterns φ_ζ^k generated from control of the plant (6.1) with different operating conditions and so with different system dynamics $f^k(x)$, while the reference orbit remains unchanged, i.e., $f_d^m(x_d) \equiv f_d(x_d)$ for all m.

In this case, the pattern-based control process consists of the following steps:

1. When the controlled system is operated under the normal condition, identify the normal system dynamics $f^k(x)$ $(k = 0)$ [or $f^0(x)$] via adaptive NN control design, and construct a normal NN controller by using the obtained control system dynamics $f_{nn}^0(x)$ as follows:

$$u^0 = -z_1 - c_2 z_2 - f_{nn}^0(x) + \dot{\alpha}_1 \qquad (6.4)$$

   This can be conducted in the same way as in Chapter 4. The above controller (6.4) will then be employed as the normal controller which can achieve specified stability and performance.

2. When the plant is controlled by the normal NN controller u^0, but the system is operated under an unusual or abnormal condition $(k \neq 0)$, that is, the system dynamics is changed to $f^k(x)$ $(k = 1, \ldots, N)$, identify the underlying system dynamics $\beta^k(x, u^0) := f^k(x) + u^0$ $(k = 1, \ldots, N)$ of training closed-loop dynamical patterns φ_ζ^k. Note that in this case, the system is still controlled by the normal NN controller u^0, which may not achieve the specified performance. The identification of $\beta^k(x, u^0)$ $(k = 1, \ldots, N)$ can be conducted in the same way as in Chapter 3. The identified training closed-loop patterns are represented by the locally accurate NN approximation $\beta_{nn}^k(x, u^0)$ $(k = 1, \ldots, N)$.

3. In the case of an abnormal condition, restart adaptive NN control design to identify the abnormal system dynamics $f^k(x)$ $(k = 1, \ldots, N)$ with guaranteed stability and tracking performance. This can be conducted in the same way as in Chapter 4. We construct a set of pattern-based NN controllers by using the obtained control system dynamics as follows:

$$u^k = -z_1 - c_2 z_2 - f_{nn}^k(x) + \dot{\alpha}_1 \qquad (6.5)$$

   where $k = 1, \ldots, N$.

4. In the recognition phase, construct dynamic models using $\beta_{nn}^k(x, u^0)$ $(k = 1, \ldots, N)$ and rapidly recognize a test closed-loop dynamical pattern. This can be conducted in the same way as in Chapter 5. The estimator with the smallest estimation error corresponds to the training closed-loop dynamical pattern which is most similar to the test closed-loop dynamical pattern.

5. Select the corresponding NN controller u^k based on the result of rapid recognition. This NN controller will be able to achieve guaranteed stability and improved control performance.

REMARK 6.2

It is seen that due to the presence of control u, identification and recognition of closed-loop dynamical patterns are more involved. The difficulty lies in how to deal with the control input u and how to construct estimators as in Chapter 5. In the above steps 2 and 4, identification and rapid recognition of closed-loop dynamical patterns are processed under normal control u^0 which is designed for normal system dynamics $f^0(x)$. Extension of this work to more general systems requires more study.

6.3   Learning Control Using Experiences

In this section, we show that when the control situation (or dynamical pattern) is correctly classified, the selected NN learning controller with knowledge or experience is able to achieve guaranteed stability and improved control performance. It is shown that with appropriate initial conditions, the NN learning controller can achieve small tracking errors and fast convergence rate with small control gains. Furthermore, the NN learning controller does not need adaptation of neural weights; the NN controller is a low-order static controller that can be more easily implemented. Thus, not only stability of the closed-loop system is guaranteed, better performance is also achieved in the aspects of time saving or energy saving. This demonstrates the benefits of knowledge utilization in control processes.

6.3.1   Problem Formulation

Consider the following nonlinear system:

$$\begin{cases} \dot{x}_1 = x_2 \\ \dot{x}_2 = f'(x) + u \end{cases} \tag{6.6}$$

where $f'(x)$ is the unknown smooth nonlinearity. Assume that system (6.6) is similar to system (6.1) in the sense that $\max_{x \in \Omega_\varsigma} |f'(x) - f^k(x)| < \varepsilon_k^*$, where Ω_ς is a compact set of interest.

Consider the following reference model:

$$\begin{cases} \dot{x}_{d_1} = x_{d_2} \\ \dot{x}_{d_2} = f'_d(x_d) \end{cases} \tag{6.7}$$

which generates a reference dynamical pattern φ_{d_ς} similar to one reference dynamical pattern φ_d^m generated from the reference model (6.2).

The control situations (either the reference dynamical pattern or the closed-loop dynamical pattern) can be recognized and classified as in Section 6.2. The objective of this section is to select an NN learning controller

$$u = -z_1 - c_2 z_2 - \overline{W}^T S(x) + \dot{\alpha}_1 \tag{6.8}$$

where z_1, z_2, α_1; and $\dot{\alpha}_1$ are given in Equations (4.4) to (4.7); $\overline{W}^T S(x)$ is the RBF approximation to the control system dynamics $f^k(x)$ obtained from the identification phases as in Section 6.2, such that (i) all the signals in the closed-loop system remain bounded, and the state tracking error $\tilde{x} = x - x_d$ converges exponentially to an arbitrarily small neighborhood of zero. (ii) Improved control performance is obtained with smaller control gains, compared with the adaptive NN control approach (4.3) and (4.8). The performance is also compared with that for a controller without using NN; that is,

$$u = -z_1 - c_2 z_2 + \dot{\alpha}_1 \tag{6.9}$$

REMARK 6.3
For the control objective (i), we do not try to achieve global or semiglobal stability of the closed-loop system. Instead, the state tracking can only be achieved for initial conditions started from the local region (as stated in Equation [4.38]), within which the NN approximation of $f(x)$ can be guaranteed.

6.3.2 Neural Network Learning Control

The following theorem shows the stability and control performance of the closed-loop system.

THEOREM 6.1
Consider the closed-loop system consisting of the plant (6.6), the reference model (6.7), and the neural learning controller (6.8) with the neural weights $\overline{W}$ being given by Equation (4.9). For initial condition $x_d(0)$ which generates the recurrent reference orbit (or the reference dynamical pattern) φ_{d_ς}, and with corresponding initial condition $x(0)$ in a close vicinity of φ_{d_ς}, we have that all signals in the closed-loop system remain bounded, and the state tracking error $\tilde{x}(t) = x(t) - x_d(t)$ converges exponentially to a small neighborhood around zero.

PROOF The derivatives of z_1 and z_2 are given as below.

$$\dot{z}_1 = \dot{x}_1 - \dot{x}_{d_1} = x_2 - x_{d_2} = -c_1 z_1 + z_2 \tag{6.10}$$

$$\dot{z}_2 = f'(x) + u - \dot{\alpha}_1 = -z_1 - c_2 z_2 - \overline{W}^T S(x) + f'(x) \tag{6.11}$$

Consider the following Lyapunov function candidate:

$$V_z = \frac{1}{2}z_1^2 + \frac{1}{2}z_2^2 \tag{6.12}$$

The derivative of V_z is

$$\dot{V}_z = z_1\dot{z}_1 + z_2\dot{z}_2$$

$$= -c_1z_1^2 - c_2z_2^2 - z_2(\overline{W}^T S(x) - f'(x))$$

Because

$$-\frac{1}{2}c_2z_2^2 - z_2(\overline{W}^T S(x) - f'(x)) \leq \frac{|\overline{W}^T S(x) - f'(x)|^2}{2c_2} \tag{6.13}$$

we have

$$\dot{V}_z \leq -c_1z_1^2 - \frac{1}{2}c_2z_2^2 + \frac{|\overline{W}^T S(x) - f'(x)|^2}{2c_2} \tag{6.14}$$

Because $x_1 - x_{d_1} = z_1$, $x_2 - x_{d_2} = z_2 - c_1z_1$, for all $\|x(t) - x_d(t)\| < d$, there exists $d_1 > 0$ (with $\|d\| - |d_1\|$ small) such that $\|z\| < d_1$, where $z = [z_1, z_2]^T$.

Using (i) the local knowledge stored in $\overline{W}$ corresponding to the training reference dynamical pattern φ_d^m and the control system dynamics $f^k(x)$, that is,

$$\text{dist}(x, \varphi_d^m) < d_m \Rightarrow |\overline{W}^T S(x) - f^k(x)| < \epsilon_k^* \tag{6.15}$$

(ii) the test reference dynamical pattern φ_{d_ς} is similar with one training reference dynamical pattern φ_d^m which implies $\text{dist}(\varphi_{d_\varsigma}, \varphi_d^m) < d_{\varsigma m}$ and $\text{dist}(x, \varphi_{d_\varsigma}) < d$, and (iii) the test control system dynamics is similar with one training control system dynamics $f^k(x)$ in the sense that

$$\max_{x \in \Omega_\varsigma} |f'(x) - f^k(x)| < \varepsilon_k^* \tag{6.16}$$

we have

$$\dot{V}_z < -c_1z_1^2 - \frac{1}{2}c_2z_2^2 + \frac{\epsilon_k^{*2} + \varepsilon_k^{*2}}{2c_2} \tag{6.17}$$

holds in a local region when $\|z\| < d_1$. Choose $c_1 \leq \frac{1}{2}c_2$. Denote

$$\delta := \frac{\epsilon_k^{*2} + \varepsilon_k^{*2}}{2c_2} \tag{6.18}$$

$$\rho := \delta/2c_1 = \frac{\epsilon_k^{*2} + \varepsilon_k^{*2}}{4c_1c_2} \tag{6.19}$$

Then Equation (6.17) satisfies

$$0 \leq V_z(t) < \rho + (V_z(0) - \rho)\exp(-2c_1t) \tag{6.20}$$

From Equation (6.20), we have

$$\sum_{k=1}^{2} \frac{1}{2} z_k^2 < \rho + (V_z(0) - \rho)\exp(-2c_1 t)$$

$$< \rho + V_z(0)\exp(-2c_1 t) \tag{6.21}$$

That is,

$$\sum_{k=1}^{2} z_k^2 < 2\rho + 2V_z(0)\exp(-2c_1 t) \tag{6.22}$$

Since ϵ_k^* is a small value thanks to the previous accurate learning as described in Section 4.2, ε_k^* is small by definition; $\rho = \frac{\epsilon_k^{*2} + \varepsilon_k^{*2}}{4c_1 c_2}$ can be made very small without high control gains c_1 and c_2. Thus, for initial condition $x_d(0)$ which generates the test reference pattern φ_{d_z}, and with initial condition $x(0)$ satisfying

$$z(0) = [x(0) - x_d(0)] \in \Omega_{z_0} := \left\{ z | V_z < \frac{1}{2} d_1^2 - \rho \right\} \tag{6.23}$$

we have

$$z(t) \in \Omega_z := \left\{ z | V_z < \frac{1}{2} d_1^2 \right\} \tag{6.24}$$

which guarantees that $\|z(t)\| < d_1$ and thus $\|x(t) - x_d(t)\| < d$. Thus, the state x will remain bounded in the local region described by Equation (6.15), in which the past experience is valid for use. Using Equation (6.8), in which $\dot{\alpha}_1$ is bounded because every term in Equation (4.7) is bounded, and $S(x)$ is bounded for all values of the NN input $Z = x$, we conclude that control u is also bounded. Thus, all the signals in the closed-loop system remain bounded.

Moreover, from Equation (6.22), given

$$\mu > \sqrt{2\rho} = \sqrt{\frac{\epsilon_k^{*2} + \varepsilon_k^{*2}}{2c_1 c_2}} \tag{6.25}$$

there exists a finite time T, determined by c_1, c_2, ϵ_k^*, and ε_k^*, such that for all $t \geq T$, $z(t)$ will converge to $\|z(t)\| < \mu$. Then, both z_1 and z_2 satisfy $|z_i(t)| < \mu$, $i = 1, 2$. Because $z_1 = x_1 - x_{d_1}$, we know that x_1 will track closely to x_{d_1}. From $z_2 = x_2 - \alpha_1 = x_2 + c_1 z_1 - x_{d_2}$, we get

$$x_2 - x_{d_2} = z_2 - c_1 z_1 \leq \mu + c_1 \mu \tag{6.26}$$

which is also a small value because μ can be made small without choosing large c_1 and c_2. Therefore, both x_1 and x_2 will exponentially converge to x_{d_1} and x_{d_2} in finite time T. This ends the proof. ∎

REMARK 6.4

From Equation (6.23), it is required that $\frac{1}{2}d_1^2 - \rho > 0$, that is, $c_1 c_2 > \frac{\epsilon_k^{*2} + \varepsilon_k^{*2}}{4d_1^2}$. Because ϵ_k^{*2} and ε_k^* are small, $c_1 c_2$ does not need to be very large with an appropriate d_1. As d_1 (and d in Equation [4.38]) represents the valid region of accurate approximation, it is seen from Equation (6.23) that the larger the d_1 (and d), the easier the selection of initial conditions.

REMARK 6.5

The larger region of operation also means better generalization ability, which is an important characteristic of neural networks, but is seldom considered in conventional NN control design [46,237]. Generalization is referred to as the ability of neural networks to provide meaningful outputs when the NN inputs are not necessarily in the training set. When using NNs in closed-loop control systems, the training examples are actually constrained by the system dynamics of both the plant and the reference model [44]. Therefore, the training set cannot be selected freely, and it often remains within a small region of the entire state space.

To expand the operation region and to improve the NN generalization, it is feasible, in the learning (or training) stage, to either inject bounded artificial noise (called jitter) [89], or track to quasi-periodic or even chaotic reference trajectories [237]. For using artificial noise, the amount of jitter needs to be carefully determined, as too much of it will obviously produce garbage, and too little of it will not have much effect [89]. Moreover, training with jitter might also damage the stability of the closed-loop system, if it is not appropriately handled. On the other hand, when using chaotic trajectories [237] to improve NN generalization, the partial PE condition and locally accurate learning can still be achieved. Further investigation will be conducted on these topics.

6.3.3   Improved Control Performance

We now compare the control performances of (i) the adaptive NN control approach (4.3), (4.8); (ii) the controller without using NN (6.9); and (iii) the neural learning control scheme (6.8). In the adaptive NN control approach (4.3), the term corresponding to δ in the above proof (see Equation [4.21]) becomes

$$\delta = \frac{\widetilde{W}^{*2} s^{*2}}{4\bar{c}_{22}} + \frac{\epsilon^{*2}}{4\bar{c}_{22}} \tag{6.27}$$

when the Lyapunov function candidate is chosen as $V_z = \frac{1}{2}z_1^2 + \frac{1}{2}z_2^2$. Alternatively, δ is in the form (see, e.g., [64])

$$\delta := \frac{\sigma \| W^* \|^2}{2} + \frac{\epsilon^{*2}}{4c_{22}} \tag{6.28}$$

when the Lyapunov function is $V = \frac{1}{2}z_1^2 + \frac{1}{2}z_2^2 + \frac{1}{2}\widetilde{W}^T \Gamma^{-1} \widetilde{W}$.

In the first case, when no result on the convergence of $\widetilde{W} = \widehat{W} - W^*$ is obtained (as in conventional adaptive NN control), δ in Equation (6.27) may be a very large value due to the possibly large $\bar{w}^*$. To keep the tracking convergence of $\tilde{x} = x - x_d$ to a small neighborhood of zero, the control gains c_1 and especially c_2 need to be chosen large enough to make δ small. In the second case, δ in Equation (6.28) can be made small by choosing a small σ and a large c_{22}; however, we cannot obtain an exponential convergence result from the chosen Lyapunov function V. In this case, only convergence of $\tilde{x} = x - x_d$ can be guaranteed as time goes to infinity.

For a controller without using NN, that is, Equation (6.9), by using the Lyapunov function candidate $V_z = \frac{1}{2}z_1^2 + \frac{1}{2}z_2^2$, the derivative of V_z is

$$\dot{V}_z = z_1\dot{z}_1 + z_2\dot{z}_2$$

$$= -c_1z_1^2 - c_2z_2^2 + z_2 f'(x)$$

$$\leq -c_1z_1^2 - \frac{1}{2}c_2z_2^2 + \frac{f^{*2}}{2c_2} \qquad (6.29)$$

where f^* denotes the upper bound of $f'(x)$ in a local region. As with the case in Equation (6.27), very high control gains are required to achieve local stability.

By comparison, using the neural learning controller (6.8), we achieved better control performance. Specifically, it can be noted: (i) smaller control gains are employed, because δ in Equation (6.18) is only related to the small constants ϵ_k^* and ε_k^*; (ii) faster tracking convergence rate is obtained, because exponential convergence is guaranteed as shown from (6.20); and (iii) smaller tracking errors can be achieved because the tracking error $\tilde{x} = x - x_d$ is related to μ which can be made very small without using high gains.

From the point of view of practical implementations, the adaptive NN control approach (4.3), (4.8) actually requires a large number of neural weights to be updated simultaneously. This makes the algorithm either energy consuming with analog hardware implementation, or time consuming with digital implementation. On the contrary, the neural learning control scheme (6.8) does not need any parameter adaptation, and can be more easily designed with both analog and digital implementations. Therefore, better performance is also achieved in the aspects of time saving or energy saving, which might be important for particular practical applications.

6.4 Simulation Studies

To demonstrate the pattern-based control approach, we again take the van der Pol oscillator (4.39) as the plant, and the Duffing oscillator (4.40) as the reference model. In Chapter 4, the van der Pol oscillator system dynamics is

$f(x_1, x_2) = -x_1 + \beta(1 - x_1^2)x_2$ where the system parameter is $\beta = 0.7$. The system dynamics of the van der Pol oscillator can be accurately approximated along the tracking trajectories. The learned system dynamics are stored in the Gaussian RBF network $\overline{W}^T S(Z)$, which can provide locally accurate NN approximation of the unknown system dynamics $f(x)$, as seen from Figures 4.2f, 4.3f, and 4.4f. Using the learning system dynamics, the corresponding NN learning controller can be constructed as (6.8).

The Duffing oscillator has been used in Chapter 4 to generate the periodic and chaotic reference orbits (as shown in Figures 4.3a and 4.4a). These reference orbits are referred to as training dynamical patterns in Chapter 5. Particularly, training pattern φ_ζ^1 is generated with initial condition $x(0) = [x_1(0), x_2(0)]^T = [0.0, -1.8]^T$, and system parameters $p_1 = 0.55$, $p_2 = -1.1$, $p_3 = 1.0$, $w = 1.8$, and $q = 1.498$. Training pattern φ_ζ^2 is generated with the same system parameters except $p_1 = 0.35$. The locally accurate NN approximation of the underlying system dynamics along the orbit of the two training patterns φ_ζ^1 and φ_ζ^2 is shown in Figures 5.2d and e and 5.3d and e. Figures 5.2f and 5.3f show the time-invariant representations of the two training patterns φ_ζ^1 and φ_ζ^2.

Moreover, we have shown in Chapter 5 that rapid recognition of test dynamical patterns can be achieved via state synchronization. Two periodic patterns, as shown in Figure 5.4, are used as the test reference dynamical patterns φ_ς^1 and φ_ς^2. Test pattern φ_ς^1 is generated from the Duffing oscillator (5.4), with initial condition $x(0) = [x_1(0), x_2(0)]^T = [0.0, -1.8]^T$ and system parameters $p_1 = 0.6$, $p_2 = -1.1$, $p_3 = 1.0$, $w = 1.8$, and $q = 1.498$. The initial condition and system parameters of test pattern φ_ς^2 are the same as those of test pattern φ_ς^1, except that $p_1 = 0.4$. From Figure 5.5, we show that test pattern φ_ς^1 is very similar to training pattern φ_ζ^1 and similar to training pattern φ_ζ^2. Test pattern φ_ς^2 is similar to the chaotic training pattern φ_ζ^2 and not very similar to the periodic training pattern φ_ζ^1, as shown in Figure 5.6.

In this section, we again use the van del Pol oscillator (4.39) and the Duffing oscillator (4.40). For the van del Pol oscillator, the system parameter β in system dynamics $f(x_1, x_2) = -x_1 + \beta(1 - x_1^2)x_2$ is changed from $\beta = 0.7$ in Chapter 4 to $\beta = 0.65$ in this chapter. The Duffing oscillator (4.40) is used to generate reference orbits, that is, the two test dynamical patterns φ_ς^1 and φ_ς^2 as in Chapter 5. The van del Pol oscillator is controlled to track the reference orbits of the Duffing oscillator by using the NN learning controller (6.8). The design parameters are $c_1 = 2$, $c_2 = 3$, which are much smaller compared with those used in Section 4.2. The initial conditions are $[x_1(0), x_2(0)]^T = [0.1, 0.2]^T$ and $[x_{d_1}(0), x_{d_2}(0)]^T = [0.2, 0.3]^T$.

First, as test pattern φ_ς^1 is recognized as very similar to training periodic pattern φ_ζ^1, we select the NN controller (6.8) based on this recognition. The NN controller (6.8) contains the learned system dynamics as experience (shown in Figure 4.3f). This experience is obtained in Chapter 4 from tracking control to the periodic orbit of the training dynamical pattern φ_ζ^1. From Figures 6.1a and b, we can see that the selected NN controller (6.8) achieves good tracking

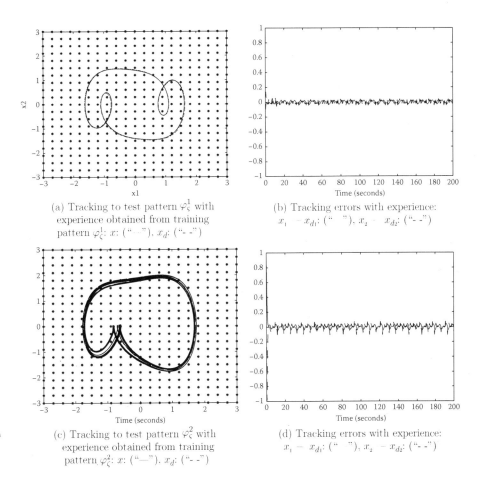

(a) Tracking to test pattern φ_ς^1 with experience obtained from training pattern φ_ζ^1: x: ("⋯"), x_d: ("- -")

(b) Tracking errors with experience: $x_1 - x_{d1}$: (" "), $x_2 - x_{d2}$: ("- -")

(c) Tracking to test pattern φ_ς^2 with experience obtained from training pattern φ_ζ^2: x: ("—"), x_d: ("- -")

(d) Tracking errors with experience: $x_1 - x_{d1}$: (" "), $x_2 - x_{d2}$: ("- -")

FIGURE 6.1
Pattern-based learning control.

to the periodic orbit of the test pattern φ_ς^1. Similarly, as test pattern φ_ς^2 is recognized as similar to training chaotic pattern φ_ζ^2, we select the NN controller (6.8) which contains the learned system dynamics as experience (shown in Figure 4.4f). From Figures 6.1c and d, we can see that the selected NN controller (6.8) achieves good tracking to the periodic orbit of the test pattern φ_ς^2.

Second, as test pattern φ_ς^1 is also recognized as similar to training chaotic pattern φ_ζ^2, we select the NN controller (6.8) which contains the learned system dynamics as experience (shown in Figure 4.4f). From Figures 6.2a and b, we can see that the selected NN controller (6.8) can still achieve good tracking to the periodic orbit of the test pattern φ_ς^1. However, when we use the NN

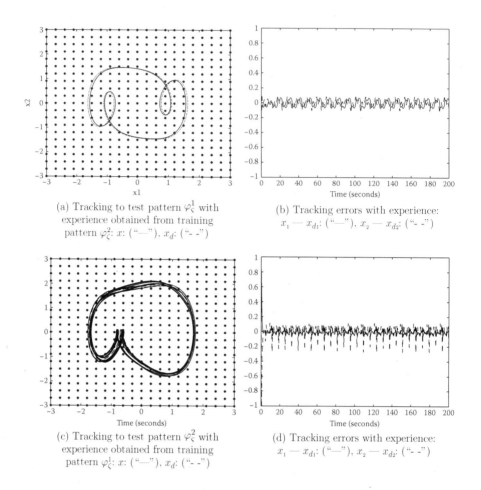

(a) Tracking to test pattern φ_ς^1 with experience obtained from training pattern φ_ς^2: x: ("—"), x_d: ("- -")

(b) Tracking errors with experience: $x_1 - x_{d_1}$: ("—"), $x_2 - x_{d_2}$: ("- -")

(c) Tracking to test pattern φ_ς^2 with experience obtained from training pattern φ_ς^1: x: ("—"), x_d: ("- -")

(d) Tracking errors with experience: $x_1 - x_{d_1}$: ("—"), $x_2 - x_{d_2}$: ("- -")

FIGURE 6.2
Pattern-based learning control.

controller (6.8) corresponding to the training periodic pattern φ_ς^1 to track to the periodic orbit of the test pattern φ_ς^2, the control performance becomes worse, as shown in Figures 6.2c and d. This implies that an NN controller trained using a chaotic dynamical pattern may be more "experienced" than one training with a periodic dynamical pattern.

If there is completely no experience, the controller without NN (6.9) is used as a comparison, which achieves much worse control performance, as shown in Figure 6.3. The simulation results clearly demonstrate how past experiences can be effectively used in pattern-based control to achieve improved performance.

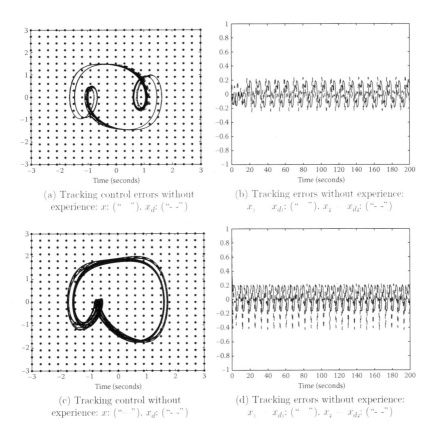

(a) Tracking control errors without experience: x: (" "), x_d: ("- -")

(b) Tracking errors without experience: $x_1 - x_{d_1}$: (" "), $x_2 - x_{d_2}$: ("- -")

(c) Tracking control without experience: x: (" "), x_d: ("- -")

(d) Tracking errors without experience: $x_1 - x_{d_1}$: (" "), $x_2 - x_{d_2}$: ("- -")

FIGURE 6.3
Tracking control without experience.

6.5 Summary

The combination of pattern recognition with control is attractive and interesting. The implementation of the idea, however, is challenging. Problems involved include: (i) learning in nonstationary or dynamic environments, (ii) representation and similarity of control situations, (iii) rapid recognition and classification of different control situations, and so on. Conventional pattern recognition methods, especially those developed for static patterns, are not suitable to cope with these problems.

In this chapter, based on the aforementioned results, we propose a new framework to implement pattern-based identification, recognition, and control in a unified way. Different control situations are defined as reference or closed-loop dynamical patterns and are identified via deterministic learning. Similar control situations can be rapidly classified, and the outcome can be used to select the suitable NN controller to achieve easy control. The proposed framework of pattern-based control bears an analogy to proficient human control with little cognitive effort. It will be useful in areas such as motion learning and control of humanoid robotics, and security assessment and control of power systems.

7

Deterministic Learning with Output Measurements

7.1 Introduction

In the preceding chapters, deterministic learning is presented for identification, recognition, and control of nonlinear systems under full-state measurements. In practice, usually there are only partial or output measurements available. It is therefore necessary to extend deterministic learning theory to such cases. The main focus of this chapter is to study knowledge acquisition, representation, and knowledge utilization in dynamic processes with output measurements.

When there are only partial or output measurements available for identification and control, it is normally required to estimate the other states of the systems. This leads to the development of linear and nonlinear observer techniques. Over the past decades, nonlinear observer design has been an active and challenging research area in the control community (see [71,169] for a survey of recent development). Early results on nonlinear observers include the Thau observer [116,220], the extended Kalman filter (EKF), and the extended Luenberger observer (ELO) [18,264]. Many attempts have been made for improvement and generalization of the ELO (see, e.g., [33,37,107,224,225,259]). One problem with the ELO is that the system dynamics are required to be (almost) exactly known. If the system dynamics are unknown, the ELO will fail to provide correct state estimation.

Another approach for nonlinear observer design is usually gathered under the category of "high-gain" observers (see, e.g., [36,61,109,111]. The design of high-gain observers aims to split the nonlinear dynamics into a linear part and a nonlinear part, and to choose the gain of the observer so that the linear part dominates the nonlinear one. By choosing the observer gain large enough, the observation error can be made arbitrarily small. However, high gains may yield large oscillations/variations in the presence of noise. It is therefore of interest to investigate how to achieve accurate state estimation in the presence of unknown system dynamics without using high gains.

Also, there are other approaches such as adaptive observers and neural networks (NN)-based observers. When the nonlinear systems contain unknown parameters, adaptive observers entail simultaneous estimation of both state variables and system parameters (see, e.g., [146,148]) provided that the PE condition is satisfied. Combined with the function approximation ability of neural networks, NN-based adaptive observers are proposed [112,192], in which neural networks are used to approximate the underlying system dynamics. Arbitrarily small state estimation error can be achieved for a class of nonlinear systems with unknown dynamics, by choosing appropriately the observer gain (usually large enough).

Although much progress has been achieved for accurate state estimation, the problem of accurate identification of the underlying system dynamics in nonlinear observer design has not been investigated in the literature. In this chapter, first, for a class of nonlinear systems undergoing periodic or recurrent motions with only output measurements, we show that locally accurate identification of nonlinear system dynamics can still be achieved. Specifically, by using a high-gain observer and a dynamical RBF network (RBFN), when state estimation is achieved by the high-gain observer, along the estimated state trajectory, a partial persistence of excitation (PE) condition is satisfied, and locally accurate identification of system dynamics is achieved in a local region along the estimated state trajectory.

Second, we show that the knowledge obtained through deterministic learning can be reused in another state observation process to achieve non-high-gain design. As high gains may yield large oscillations/variations in the presence of noise, it is not appropriate to rely on high-gain design in all situations. To achieve state estimation without using high gains, the knowledge on system dynamics is normally required to be known. Because the learned knowledge stored in the constant RBF networks actually provides locally accurate system dynamics, we naturally use the knowledge to construct an RBFN-based nonlinear observer, in which the constant RBF networks are embedded as NN approximations for system dynamics. For state estimation of the same nonlinear system as previously observed, it is shown that correct state estimation can be achieved according to the internal matching of the underlying system dynamics without using high-gain domination.

Third, we show that the results on deterministic learning with output measurements and non-high-gain observer design are applicable to effective representation and rapid recognition of single-variable dynamical patterns. Specifically, a single-variable dynamical pattern can be represented in a time-invariant and spatially distributed manner via deterministic learning and state observation. This representation is a kind of static representation. Moreover, for a set of training single-variable dynamical patterns, a set of RBFN-based observers is constructed within which the constant RBF networks are embedded. These RBFN-based observers are taken as dynamic representations for the corresponding training single-variable dynamical patterns.

For rapid recognition of a test single-variable dynamical pattern from a set of training single-variable dynamical patterns, we take the test pattern as an input to each RBFN-based observer for the corresponding training pattern. A state observation error system is yielded corresponding to the nonlinear system generating the test pattern and the RBFN-based observer. The non-high-gain observation errors are proven to be approximately proportional to the differences on system dynamics of the test and training dynamical patterns, thus they can be taken as the measure of similarity between the test and training dynamical patterns. Note that although most state variables of the test pattern are not available from measurement, a high-gain observer can be employed again to provide accurate estimates of these state variables, so that the non-high-gain observation errors can still be computed. For similar test and training dynamical patterns, the non-high-gain observation errors converge to small neighborhoods of zero due to a kind of internal matching of system dynamics of the test and training patterns. The training single-variable dynamical pattern whose corresponding observer yields the smallest observation error will be recognized as most similar to the given test single-variable dynamical pattern.

The results of this chapter draw on the recent papers [241,246,248].

7.2 Learning from State Observation

In this section, we investigate how to achieve deterministic learning from state observation for a class of nonlinear systems undergoing recurrent motions. The problem formulation is as follows.

Consider a class of nonlinear systems in the following observable form

$$
\begin{cases}
\dot{x}_1 = x_2 \\
\dot{x}_2 = x_3 \\
\quad \vdots \\
\dot{x}_{n-1} = x_n \\
\dot{x}_n = f(x) \\
y = x_1
\end{cases}
\tag{7.1}
$$

where $x = [x_1, \ldots, x_n]^T \in R^n$ is the system state, y is the system output which is measurable, and $f(x)$ is a smooth, unknown nonlinear function.

ASSUMPTION 7.1
Assume that system state $x(t)$ remains uniformly bounded; that is, $x(t) \in \Omega \subset R^n$, $\forall t \geq t_0$, where Ω is a compact set. Moreover, the system trajectory starting from initial condition x_0, denoted as $\varphi_\zeta(t, x_0)$ (or φ_ζ for conciseness), is a recurrent trajectory.

The objective is to identify the unknown dynamics $f(x)$ along the trajectory $\varphi_\zeta(t, x_0)$, by using only output measurement $y = x_1$.

The objective can be implemented in two steps. First, we use the following high-gain observer [61] to estimate the state variables $x_2, \ldots, x_n$:

$$\dot{\hat{x}}_1 = \hat{x}_2 + h_1 k(y - \hat{y})$$
$$\dot{\hat{x}}_2 = \hat{x}_3 + h_2 k^2(y - \hat{y})$$
$$\vdots \tag{7.2}$$
$$\dot{\hat{x}}_{n-1} = \hat{x}_n + h_{n-1} k^{n-1}(y - \hat{y})$$
$$\dot{\hat{x}}_n = h_n k^n(y - \hat{y})$$
$$\hat{y} = \hat{x}_1$$

where $h_i (i = 1, \ldots, n)$ and k are design constants, $\hat{x} = [\hat{x}_1, \ldots, \hat{x}_n]^T$ is the estimate of the state x, and $\hat{y}$ denotes the estimate of system output y. If h_i is chosen such that $s^n + \sum_{j=1}^{n} h_j s^{n-j}$ is a Hurwitz polynomial with distinct roots, then for all d and all times t' there exists a finite observer gain k' such that for all $k \geq k'$, the observer error satisfies $\|\hat{x}(t) - x(t)\| \leq d, \forall t \geq t'$ [61]. For convenience of presentation, this is denoted as $\hat{x} \to x$, which means that the estimates $\hat{x}(t)$ converge to a sufficiently small neighborhood of the state $x(t)$ in a finite time.

REMARK 7.1

Note that the employment of the above high-gain observer requires that $f(x)$ in Equation (7.1) be global Lipschitz [61]. Because $x(t)$ is assumed to be uniformly bounded, $x(t) \in \Omega \subset R^n$, $\forall t \geq t_0$, the global Lipschitz condition on $f(x)$ [61] is actually satisfied within the compact set Ω.

Second, we employ the following dynamical RBF network to identify the dynamics $f(x)$:

$$\dot{\chi} = -a(\chi - \hat{x}_n) + \widehat{W}^T S(\hat{x}) \tag{7.3}$$

where χ is the state of the dynamical RBF network, $\hat{x}_n$ is a state variable of observer (7.2), $a > 0$ is a design constant, and a localized RBF network $\widehat{W}^T S(\hat{x})$ is used to approximate the unknown $f(x)$. The neural weights $\widehat{W}$ are updated by

$$\dot{\widehat{W}} = \dot{\widetilde{W}} = -\Gamma S(\hat{x})\bar{x}_n - \sigma \Gamma \widehat{W} \tag{7.4}$$

where $\Gamma = \Gamma^T > 0, \sigma > 0$ is a small value, and $\bar{x}_n$ is defined as $\bar{x}_n := \chi - \hat{x}_n$. We also define $\tilde{x}_n := \chi - x_n$.

It is seen that only the estimated information $\hat{x}$, as well as χ, is used in Equations (7.3) and (7.4). The state x of system (7.1), because it is mostly not available from measurement, does not appear. Note also that $\bar{x}_n = \chi - \hat{x}_n$ is computable, whereas $\tilde{x}_n = \chi - x_n$ is not. However, it is seen that as $\hat{x}_n \to x_n$, $\tilde{x}_n \to \bar{x}_n$.

The following theorem indicates that identification of the unknown $f(x)$ can be achieved along the trajectory $\varphi_\varsigma(x_0)$ when $\hat{x} \to x$.

THEOREM 7.1

Consider the adaptive system consisting of the nonlinear dynamical system (7.1), the high-gain observer (7.2), the dynamical RBF network (7.3), and the NN weight adaptation law (7.4). For a periodic or recurrent trajectory $\varphi_\zeta(x_0)$, with initial values $\widehat{W}(0) = 0$, we have: (i) all signals in the adaptive system remain uniformly bounded; and (ii) locally accurate approximation for the unknown $f(x)$ to the error level ϵ^* is obtained along the trajectory $\varphi_\zeta(x_0)$ when $\hat{x} \to x$.

PROOF   (i) Boundedness of all the signals in the adaptive system is first analyzed. With high-gain observer (7.2), $\hat{x} \to x$ and $\tilde{x}_n \to \bar{x}_n$. From Equations (7.1) and (7.3), the derivative of $\tilde{x}_n = \chi - x_n$ satisfies

$$
\begin{aligned}
\dot{\tilde{x}}_n &= \dot{\chi} - \dot{x}_n \\
&= -a(\chi - \hat{x}_n) + \widehat{W}^T S(\hat{x}) - f(x) \\
&= -a\tilde{x}_n + \widehat{W}^T S(\hat{x}) - f(x) \\
&= -a\tilde{x}_n + \widehat{W}^T S(\hat{x}) - f(\hat{x}) + f(\hat{x}) - f(x) \\
&= -a\tilde{x}_n + a(\tilde{x}_n - \bar{x}_n) + \widehat{W}^T S(\hat{x}) - W^{*T} S(\hat{x}) - \epsilon + f(\hat{x}) - f(x) \\
&= -a\tilde{x}_n + \widetilde{W}^T S(\hat{x}) + \varepsilon
\end{aligned}
\tag{7.5}
$$

where $\widetilde{W} = \widehat{W} - W^*$, and (by combining Equation [7.3] and using the Intermediate Value Theorem [110])

$$
\begin{aligned}
\varepsilon &= a(\tilde{x}_n - \bar{x}_n) - \epsilon + f(\hat{x}) - f(x) \\
&= a(\tilde{x}_n - \bar{x}_n) - \epsilon + \left.\frac{\partial f(x)}{\partial x}\right|^T_{x=\hat{x}'}(\hat{x} - x) \\
&< a|\tilde{x}_n - \bar{x}_n| + \left\|\left.\frac{\partial f(x)}{\partial x}\right\|_{x=\hat{x}'}\right\| \|\hat{x} - x\| + \epsilon^*
\end{aligned}
\tag{7.6}
$$

in which $\hat{x}' \in [\hat{x}, x)$ or $\hat{x}' \in (x, \hat{x}]$. It is seen that when $\hat{x} \to x_j$, $\varepsilon = O(\epsilon)$ because a and $\|\frac{\partial f(x)}{\partial x}\|_{x=\hat{x}'}$ are bounded, $\tilde{x}_n - \bar{x}_n$ is small when $\hat{x} \to x$, and ϵ^* can be chosen to be small.

Again, with $\hat{x}_n \to x_n$ and $\tilde{x}_n \to \bar{x}_n$, adaptation law (7.4) is expressed by

$$
\dot{\widehat{W}} = \dot{\widetilde{W}} = -\Gamma S(\hat{x})\tilde{x}_n - \sigma\Gamma\widehat{W} - \varepsilon_W
\tag{7.7}
$$

where $|\varepsilon_W| = |\Gamma S(\hat{x})(\tilde{x}_n - \bar{x}_n)|$ is small when $\hat{x}_n \to x_n$.

Consider the following Lyapunov function candidate:

$$
V = \frac{1}{2}\tilde{x}_n^2 + \frac{1}{2}\widetilde{W}^T\Gamma^{-1}\widetilde{W}
\tag{7.8}
$$

The derivative of V along solutions of Equation (7.5) is

$$
\begin{aligned}
\dot{V} &= \tilde{x}_n\dot{\tilde{x}}_n + \widetilde{W}^T\Gamma^{-1}\dot{\widetilde{W}} \\
&= -a\tilde{x}_n^2 - \tilde{x}_n\varepsilon - \sigma\widetilde{W}^T\widehat{W} - \widetilde{W}^T S(\hat{x})(\tilde{x}_n - \bar{x}_n)
\end{aligned}
$$

Let $a = a_1 + a_2$ with $a_1, a_2 > 0$. Since

$$-a_2 \tilde{x}_n^2 - \tilde{x}_n \varepsilon \le \frac{\varepsilon^2}{4a_2}$$

$$-\sigma \widetilde{W}^T \widehat{W} - \widetilde{W}^T S(\hat{x})(\tilde{x}_n - \bar{x}_n)$$

$$\le -\sigma \|\widetilde{W}\|^2 + \sigma \|\widetilde{W}\| \|W^*\| + \|\widetilde{W}\| s^* d$$

$$\le -\frac{\sigma \|\widetilde{W}\|^2}{2} + \frac{\sigma (\|W^*\| + s^* d/\sigma)^2}{2}$$

it follows that

$$\dot{V} \le -a_1 \tilde{x}_n^2 - \frac{\sigma \|\widetilde{W}\|^2}{2} + \frac{\sigma (\|W^*\| + s^* d/\sigma)^2}{2} + \frac{\varepsilon^2}{4a_2}$$

From the above, it is clear that $\dot{V}$ is negative definite whenever $|\tilde{x}_n| > \frac{\varepsilon}{2\sqrt{a_1 a_2}} + \sqrt{\frac{\sigma}{2a_1}}(\|W^*\| + s^* d/\sigma)$, or $\|\widetilde{W}\| > \frac{\varepsilon}{\sqrt{2\sigma a_2}} + (\|W^*\| + s^* d/\sigma)$. This leads to the uniform boundedness of both $\tilde{x}_n$ and $\widetilde{W}$ as

$$|\tilde{x}_n| \le \frac{\varepsilon}{2\sqrt{a_1 a_2}} + \sqrt{\frac{\sigma}{2a_1}}(\|W^*\| + s^* d/\sigma) \tag{7.9}$$

$$\|\widetilde{W}\| \le \frac{\varepsilon}{\sqrt{2\sigma a_2}} + (\|W^*\| + s^* d/\sigma) \tag{7.10}$$

From the boundedness of $\tilde{x}_n$ and $\widetilde{W}$, we see that both χ and $\widehat{W}$ are uniformly bounded. Thus, all the signals in the adaptive system remain uniformly bounded.

(ii) By using the spatially localized learning property of RBF networks, as shown in Equation (2.12), along the estimated system trajectory $\hat{x}(t)$, the derivative of $\tilde{x}$, that is, Equation (7.5), is described by

$$\dot{\tilde{x}}_n = -a \tilde{x}_n + \widehat{W}_\zeta^T S_\zeta(\hat{x}) + \widehat{W}_{\bar{\zeta}}^T S_{\bar{\zeta}}(\hat{x})$$
$$- W_\zeta^* S_\zeta(\hat{x}) - \epsilon_\zeta + f(\hat{x}) - f(x)$$
$$= -a \tilde{x}_n + \widehat{W}_\zeta^T S_\zeta(\hat{x}) + \varepsilon_\zeta \tag{7.11}$$

where $S_\zeta(\hat{x})$ is a subvector of $S(\hat{x})$, $\widehat{W}_\zeta$ is the corresponding weight subvector, the subscript $\bar{\zeta}$ stands for the region far away from the estimated state trajectory $\hat{x}$, with $|\widehat{W}_{\bar{\zeta}}^T S_{\bar{\zeta}}(\hat{x})|$ being small, and

$$\varepsilon_\zeta = a(\tilde{x}_n - \bar{x}_n) - \epsilon_\zeta + f(\hat{x}) - f(x) + \widehat{W}_{\bar{\zeta}}^T S_{\bar{\zeta}}(\hat{x})$$
$$= a(\tilde{x}_n - \bar{x}_n) - \epsilon_\zeta + \frac{\partial f(x)}{\partial x}\Big|_{x=\hat{x}'}^T (\hat{x} - x) + \widehat{W}_{\bar{\zeta}}^T S_{\bar{\zeta}}(\hat{x})$$
$$= \varepsilon - (\epsilon_\zeta - \epsilon) + \widehat{W}_{\bar{\zeta}}^T S_{\bar{\zeta}}(\hat{x}) \tag{7.12}$$

is the NN approximation error along the trajectory $\hat{x}$, which can be expressed as $O(\epsilon)$ since $\varepsilon = O(\epsilon)$, $\epsilon_\zeta = O(\epsilon)$, $\widehat{W}_{\bar{\zeta}}$ is bounded and each element of $S_{\bar{\zeta}}(\hat{x})$ is small.

Equation (7.7) is described by

$$\dot{\hat{W}}_\zeta = \dot{\tilde{W}}_\zeta = -\Gamma_\zeta S_\zeta(\hat{x})\tilde{x}_n - \sigma\Gamma_\zeta\widehat{W}_\zeta - \varepsilon_{W_\zeta} \tag{7.13}$$

and

$$\dot{\hat{W}}_{\bar{\zeta}} = \dot{\tilde{W}}_{\bar{\zeta}} = -\Gamma_{\bar{\zeta}} S_{\bar{\zeta}}(\hat{x})\tilde{x}_n - \sigma\Gamma_{\bar{\zeta}}\widehat{W}_{\bar{\zeta}} - \varepsilon_{W_{\bar{\zeta}}} \tag{7.14}$$

where $||\varepsilon_{W_\zeta}|| = ||\Gamma_\zeta S_\zeta(\hat{x})(\tilde{x}_n - \bar{x}_n)||$ and $||\varepsilon_{W_{\bar{\zeta}}}|| = ||\Gamma_{\bar{\zeta}} S_{\bar{\zeta}}(\hat{x})(\tilde{x}_n - \bar{x}_n)||$ are small when $\hat{x}_n \to x_n$.

Thus, Equations (7.11) and (7.13) are described by

$$\begin{bmatrix} \dot{\tilde{x}}_n \\ \dot{\tilde{W}}_\zeta \end{bmatrix} = \begin{bmatrix} -a & S_\zeta(\hat{x})^T \\ -\Gamma_\zeta S_\zeta(\hat{x}) & 0 \end{bmatrix} \begin{bmatrix} \tilde{x}_n \\ \tilde{W}_\zeta \end{bmatrix} + \begin{bmatrix} -\varepsilon_\zeta \\ -\sigma\Gamma\widehat{W}_\zeta - \varepsilon_{W_\zeta} \end{bmatrix} \tag{7.15}$$

Since the system state $x(t)$ is in recurrent motion, the convergence of $\hat{x}$ to x makes $\hat{x}$ also become recurrent. Then, according to Theorem 2.7, $S_\zeta(\hat{x})$ is PE almost always. With PE of $S_\zeta(\hat{x})$, according to Theorem 2.4, the exponential stability of $(\tilde{x}, \tilde{W}_\zeta) = 0$ for the nominal part of system (7.15) is achieved. Since $\varepsilon_\zeta = O(\epsilon)$, ε_{W_ζ} is small, and $\sigma||\Gamma_\zeta\widehat{W}_\zeta||$ can be made small by choosing σ small enough, by using Theorem 2.6, the parameter error $\tilde{W}_\zeta(t) = \widehat{W}_\zeta - W_\zeta^*$ converges exponentially to a small neighborhood of zero, with the size of the neighborhood being determined by ϵ^*, $||\varepsilon_{W_\zeta}||$ and $\sigma||\Gamma_\zeta\widehat{W}_\zeta||$.

The convergence of $\widehat{W}_\zeta$ to a small neighborhood of W_ζ^* implies that along the trajectory $\hat{x}(t)$, we have

$$f(\hat{x}) = W_\zeta^{*T} S_\zeta(\hat{x}) + \epsilon_\zeta$$

$$= \widehat{W}_\zeta^T S_\zeta(\hat{x}) - \tilde{W}_\zeta^T S_\zeta(\hat{x}) + \epsilon_\zeta$$

$$= \widehat{W}_\zeta^T S_\zeta(\hat{x}) + \epsilon_{\zeta_1} \tag{7.16}$$

$$= \widehat{W}^T S(\hat{x}) + \epsilon_1 \tag{7.17}$$

where $\epsilon_{\zeta_1} = \epsilon_\zeta - \tilde{W}_\zeta^T S_\zeta(Z)$. It is clear that $\epsilon_{\zeta_1} = O(\epsilon_\zeta) = O(\epsilon)$, $\epsilon_1 = O(\epsilon_{\zeta_1}) = O(\epsilon)$. Thus, it can be concluded that the entire RBF network $\widehat{W}^T S(\hat{x})$ can approximate the unknown $f(\hat{x})$ along the trajectory $\hat{x}(t)$.

Moreover, by choosing $\overline{W}$ as in (3.15), i.e.,

$$\overline{W} = \text{mean}_{t\in[t_a,t_b]} \widehat{W}(t) \tag{7.18}$$

where "mean" is the arithmetic mean [39], and $t_b > t_a > 0$ represents a time segment after the transient process, the system dynamics $f(\hat{x})$ can be described using constant RBF networks as

$$f(\hat{x}) = \overline{W}^T S(\hat{x}) + \epsilon_2 \tag{7.19}$$

where $\epsilon_2 = O(\epsilon_1) = O(\epsilon)$. Thus, with only output measurement $y = x_1$, locally accurate identification of system dynamics $f(x)$ to the error level ϵ is achieved along the trajectory $\varphi_\zeta(x_0)$ when $\hat{x} \to x$. ∎

The local region (denoted by $\Omega_{\varphi_{\zeta_y}}$) can be described by:

$$\Omega_{\varphi_{\zeta_y}} := \left\{ Z \mid \text{dist}(Z, \varphi_{\zeta_y}) < d_y \Rightarrow \left| \overline{W}^T S(Z) - f(Z; p) \right| < \epsilon_2^* \right\} \quad (7.20)$$

where d_y is a constant representing the size of the local region (d_y can be made larger via appropriate training), and the maximum approximation error ϵ_2^* is close to ϵ^*.

REMARK 7.2
The system (7.1) is very simple in form, but there appears to be few results in the literature that achieve the identification of system dynamics in a nonlinear observer problem. The result in this section can be extended to more general nonlinear systems with disturbances/noise for which high-gain observers have been successfully designed.

7.3   Non-High-Gain Observer Design

In this section, we show that the knowledge obtained through deterministic learning can be reused in another state observation process to achieve non-high-gain design.

In the literature on observer design, it is known that high gains may yield large oscillations/variations in the presence of noise. Therefore, it is useful if non-high-gain state observation can be achieved in as many situations as possible. To achieve non-high-gain estimation of state variables $x_2, \ldots, x_n$ using the output $y = x_1$ of system (7.1), knowledge of system dynamics $f(x)$ is normally required. We notice that the learned knowledge (7.20) stored in the constant RBF network actually provides locally accurately known system dynamics. For state observation of the same nonlinear system (7.27), an RBFN-based nonlinear observer is constructed as follows:

$$\dot{\hat{x}}_1 = \hat{x}_2 + k_1(y - \hat{y})$$
$$\dot{\hat{x}}_2 = \hat{x}_3 + k_2(y - \hat{y})$$
$$\vdots \qquad\qquad\qquad (7.21)$$
$$\dot{\hat{x}}_{n-1} = \hat{x}_n + k_{n-1}(y - \hat{y})$$
$$\dot{\hat{x}}_n = \overline{W}^T S(\hat{x}) + k_n(y - \hat{y})$$
$$\hat{y} = \hat{x}_1$$

where $K = [k_1, \ldots, k_n]^T$ are observer gains, $\hat{x} = [\hat{x}_1, \ldots, \hat{x}_n]^T$ are estimates of the state x, $\hat{y}$ denote the estimate of system output y, and the constant RBF network $\overline{W}^T S(\hat{x})$ provides a locally-accurate approximation of the system dynamics $f(x)$.

Define $e = x - \hat{x}$ and $\tilde{y} = y - \hat{y}$. The error dynamics of state observation is derived from Equations (7.1) and (7.21):

$$\dot{e} = (A - KC^T)e + B[f(x) - \overline{W}^T S(\hat{x})]$$
$$= (A - KC^T)e + B[f(x) - f(\hat{x})] + B[f(\hat{x}) - \overline{W}^T S(\hat{x})] \qquad (7.22)$$

where

$$A = \begin{bmatrix} 0 & 1 & 0 & \cdots & 0 \\ 0 & 0 & 1 & \cdots & 0 \\ & & \vdots & & \\ 0 & 0 & 0 & 1 & 0 \\ 0 & 0 & 0 & \cdots & 0 \end{bmatrix}, \quad B = \begin{bmatrix} 0 \\ 0 \\ \vdots \\ 0 \\ 1 \end{bmatrix}, \quad C = \begin{bmatrix} 1 \\ 0 \\ \vdots \\ 0 \\ 0 \end{bmatrix}$$

Since $x(t)$ is bounded, the stability and convergence analysis for the RBFN-based observer (7.22) can be conducted in a similar way to the analysis for Lipschitz nonlinear observers, for example, [188]. Specifically, necessary and sufficient conditions on the stability matrix that ensure asymptotic stability of the Lipschitz nonlinear observer are presented [188]. These conditions are then reformulated to obtain a sufficient condition for stability in terms of the eigenvalues and the eigenvectors of the linear stability matrix.

The following theorem shows that the RBFN-based nonlinear observer (7.21) can achieve non-high-gain state observation.

THEOREM 7.2
Assume that $\hat{x}(0) \in \Omega_{\varphi_\zeta}$. If the observer gain K is chosen such that the matrix $(A - KC^T)$ is stable and all the eigenvalues λ of $(A - KC^T)$ satisfy

$$Re(-\lambda) > K_2(T)\gamma \qquad (7.23)$$

where $(A - KC^T) = T\Lambda T^{-1}$, $K_2(T)$ is the condition number (l_2 norm) of the matrix T, and $\gamma = \max_{x \in \Omega_\Delta} \|\frac{\partial f(x)}{\partial x}\|$. Then, the state estimation error will asymptotically converge to a small neighborhood of zero without using high-gain domination.

PROOF For the error dynamics (7.22), consider Lyapunov function $V = e^T P e$. Its derivative satisfies

$$\dot{V} = e^T[(A - KC^T)^T P + P(A - KC^T)]e$$
$$+ 2e^T P[B(f(x) - f(\hat{x}))] + 2e^T P[B(f(\hat{x}) - \overline{W}^T S(\hat{x}))]$$

Since $(A - KC^T)$ is stable and Equation (7.23) holds, then according to [188, Theorem 5], the following inequality holds [188]:

$$\min_{\omega \in R^+} \sigma_{min}(A - KC - j\omega I) > \gamma \qquad (7.24)$$

According to [188, Theorem 3], there exists a symmetric positive definite matrix P and a constant $\varpi > 0$ such that

$$(A - KC^T)^T P + P(A - KC^T) + \gamma^2 PP + I + \varpi I = 0 \qquad (7.25)$$

Then, we have

$$
\begin{aligned}
\dot{V} &\le e^T[(A - KC^T)^T P + P(A - KC^T) + \gamma^2 PP + I]e + 2\|PB\|\,\|e\|\epsilon_2 \\
&\le -\varpi\|e\|^2 + 2\|PB\|\,\|e\|\epsilon_2^* \\
&\le -\frac{1}{2}\varpi\|e\|^2 + \frac{2(\|PB\|\epsilon_2^*)^2}{\varpi} \qquad (7.26)
\end{aligned}
$$

If $\|e\| > \frac{2\|PB\|\epsilon_2}{\varpi}$, then $\dot{V} < 0$. Thus, boundedness of e can be guaranteed. Since $x(t)$ is bounded, it follows that $\hat{x}$ is bounded. As ϵ_2^* is the approximation error given in Equation (7.19), which can be made small, it is concluded that the estimation error will asymptotically converge to a small neighborhood of zero, that is, $\|e\| \le e_M = \frac{2\|PB\|\epsilon_2^*}{\varpi}$. ∎

REMARK 7.3
It is clear that the size of the neighborhood can be made small not by using high-gain domination, but by choosing the observer gain K such that $(A - KC^T)$ is stable and the eigenvectors are well-conditioned. A systematic computational algorithm for choosing the observer gain is given in [188].

The results for deterministic learning (DL) applied to state observation and the RBFN-based observer together provide a new approach to observer design for uncertain nonlinear systems. The whole DL-based approach makes use of different observer design techniques: first, with the employment of a high-gain observer and an adaptive NN adaptation law, locally accurate identification of the unknown system dynamics is achieved. Second, by using the Lipschitz nonlinear observer design, the RBFN-based observer can achieve correct state estimation for the same (and also similar) nonlinear system without using high gains. The DL-based approach does not require that the system dynamics $f(x)$ is given *a priori*. Instead, knowledge learned on system dynamics from previous state observation is reused to provide an approximation to $f(x)$, such that the RBFN-based observer does not require high-gain domination. Therefore, fundamental knowledge is acquired and utilized in the state observation processes, and the disadvantages caused by high-gain design can be finally overcome, which can be regarded as the benefit of knowledge utilization in observer design.

7.4 Rapid Recognition of Single-Variable Dynamical Patterns

In this section, we consider the problem of representation, similarity definition, and rapid recognition of single-variable dynamical patterns.

The definition of a *single-variable dynamical pattern* is given as follows.

DEFINITION 7.1

A single-variable dynamical pattern is defined as a recurrent system output trajectory $y_\zeta(t)$ generated from the following nonlinear dynamical system:

$$\begin{cases} \dot{x}_{\zeta 1} = x_{\zeta 2} \\ \dot{x}_{\zeta 2} = x_{\zeta 3} \\ \quad \vdots \\ \dot{x}_{\zeta n-1} = x_{\zeta n} \\ \dot{x}_{\zeta n} = f_\zeta(x_\zeta) \\ y_\zeta = x_{\zeta 1} \end{cases} \tag{7.27}$$

where $x_\zeta = [x_{\zeta 1}, \dots, x_{\zeta n}]^T \in R^n$ is the system state, $f_\zeta(x_\zeta)$ is a smooth, unknown nonlinear function, and $y_\zeta(t)$ is the measurable system output trajectory. The single-variable dynamical pattern is denoted as φ_{ζ_y} for concise presentation.

The general recognition process for single-variable dynamical patterns still consists of the identification phase and the recognition phase as described in Chapter 5. By using deterministic learning theory and state observation techniques, the identification of a single-variable dynamical pattern is conducted in the same way as in Section 7.2. Accordingly, the dynamics of single-variable dynamical patterns can be accurately identified and stored in constant RBF networks.

For representation, similarity definition, and rapid recognition of single-variable dynamical patterns, difficulties arise not only because dynamical patterns evolve with time, but also due to the incomplete information available. In Subsections 7.4.1 and 7.4.2, we address the problems of how to appropriately represent the single-variable dynamical patterns and how to measure the similarity between two single-variable dynamical patterns, respectively. Rapid recognition of single-variable dynamical patterns via non-high-gain observation will be studied in Subsection 7.4.3.

7.4.1 Representation Using Estimated States

For representation of a single-variable dynamical pattern, complete information on both its estimated pattern states and its underlying system dynamics is used. The representation in the form of constant RBF networks can be taken as a static representation for a single-variable dynamical pattern. An RBFN-based observer with the constant RBF networks embedded is taken as a dynamic representation for the corresponding training dynamical pattern.

We have the following statements concerning the representation of a single-variable dynamical pattern:

1. A single-variable dynamical pattern φ_{ζ_y} can be represented via deterministic learning by using the constant RBF network $\overline{W}^T S(Z)$, which provides a locally accurate NN approximation of the underlying system dynamics $f_\zeta(x_\zeta)$. The knowledge represented in RBF network $\overline{W}^T S(Z)$ is valid in a local region $\Omega_{\varphi_{\zeta_y}}$, which can be described as: for the pattern state trajectory φ_{ζ_y}, there exist constants $d_y, \xi_y^* > 0$, such that

$$\text{dist}(Z, \varphi_{\zeta_y}) < d_y \Rightarrow \left| \overline{W}^T S(Z) - f_\zeta(x_\zeta) \right| < \xi_y^* \qquad (7.28)$$

where ξ_y^* is the approximation error within $\Omega_{\varphi_{\zeta_y}}$ which is also small.

2. The representation of a single-variable dynamical pattern is time-invariant because in $\overline{W}^T S(Z)$ the time attribute is eliminated. The representation is also *spatially distributed* in the sense that relevant information is stored in a large number of neurons distributed along the estimated state trajectory. Thus, a single-variable dynamical pattern is represented in a time-invariant and spatially distributed manner by using information regarding both its estimated pattern states $\hat{x}_\zeta$ and its underlying system dynamics $f_\zeta(x_\zeta)$ along the estimated state trajectory $\hat{x}_\zeta(t)$. The time-invariant and spatially distributed representation can be considered as a kind of graph-based representation. It may not be appropriate to represent a single-variable dynamical pattern by using only a limited number of features extracted from the time-varying dynamical patterns.

3. After a training single-variable dynamical pattern φ_{ζ_y} is represented using the constant RBF network $\overline{W}^T S(Z)$, an RBFN-based dynamical model is constructed within which the constant RBF network is embedded. This RBFN-based dynamical model, as introduced later, is a nonlinear observer and is taken as a dynamical representative for the corresponding training dynamical pattern φ_{ζ_y}. Thus, the complete representation of a single-variable dynamical pattern consists of two parts, the static part described by constant RBF network $\overline{W}^T S(Z)$, and the dynamical part described by the RBFN-based nonlinear observer. The static representation is suitable for storage of a time-varying dynamical pattern, however, it will not be used for rapid recognition of other dynamical patterns. Instead, the dynamical model will be used to achieve rapid recognition of dynamical patterns via non-high-gain state observation.

7.4.2 Similarity Definition

We extend the similarity definitions for full-state dynamical patterns as proposed in Chapter 5 to single-variable dynamical patterns.

Consider the dynamical pattern φ_{ζ_y} (as given by Equation [7.27]), and another dynamical pattern (denoted as φ_{ς_y}) generated from the following nonlinear dynamical system:

$$\begin{cases} \dot{x}_{\varsigma 1} = x_{\varsigma 2} \\ \dot{x}_{\varsigma 2} = x_{\varsigma 3} \\ \quad \vdots \\ \dot{x}_{\varsigma n-1} = x_{\varsigma n} \\ \dot{x}_{\varsigma n} = f_\varsigma(x_\varsigma) \\ y_\varsigma = x_{\varsigma 1} \end{cases} \tag{7.29}$$

where $x_\varsigma = [x_{\varsigma 1}, \ldots, x_{\varsigma n}]^T \in R^n$ is the state variable of the test dynamical pattern, $f_\varsigma(x_\varsigma)$ is a smooth, unknown nonlinear function, and $y_\varsigma(t)$ is the measurable output variable of the test dynamical pattern. It is assumed that $x_\varsigma(t)$ remains bounded for all time; that is, $x_\varsigma(t) \in \Omega_\Delta$, $\forall t \geq 0$.

Since the state variables are mostly unknown, it is inconvenient to characterize the similarity of single-variable dynamical patterns using the difference between system dynamics along the orbit of the test pattern, as in Definitions (5.1) and (5.2). Instead, we rely on the difference between corresponding system dynamics within a local region Ω_ς along the orbit of the test pattern

$$\Omega_\varsigma := \{x \mid \text{dist}(x, \varphi_{\varsigma_y}) < d_y\}$$

where $d_y > 0$ is a constant.

We have the following definitions for similarity of single-variable dynamical patterns.

DEFINITION 7.2

Dynamical pattern φ_{ς_y} (given by Equation [7.29]) is said to be similar to dynamical pattern φ_{ζ_y} (given by Equation [7.27]), if the state of pattern φ_{ς_y} stays within a neighborhood region of the state of pattern φ_{ζ_y}, and the difference between the corresponding system dynamics within a local region Ω_ς, that is, $\Delta f_y = |f_\zeta(x) - f_\varsigma(x)|_{\forall x \in \Omega_\varsigma} \leq \varepsilon_y^*$, where $\varepsilon_y^* > 0$ is the similarity measure, is small.

Since only a single-variable state of a dynamical pattern is available, the above definition cannot be used directly for recognition. Based on deterministic learning and state observation, the following similarity definition is given for practical use.

DEFINITION 7.3

Dynamical pattern φ_{ς_y} (given by Equation [7.29]) is recognized to be similar to dynamical pattern φ_{ζ_y} (given by Equation [7.27]), if the state of pattern φ_{ς_y} stays within a neighborhood region of the state of pattern φ_{ζ_y}, and the difference between the corresponding system dynamics within a local region Ω_{ς}, that is, $\triangle f_{Ny} = |\overline{W}^T S(x) - f_{\varsigma}(x)|_{\forall x \in \Omega_{\varsigma}} \leq \varepsilon_y^* + \xi_y^*$, where ε_y^* is the similarity measure and ξ_y^* is the approximation error given in Equation (7.28), is small.

7.4.3   Rapid Recognition via Non-High-Gain State Observation

In this subsection, we present how to achieve rapid recognition of single-variable dynamical patterns via non-high-gain state observation.

Consider a single-variable dynamical pattern φ_{ς_y} (as given by Equation (7.29)) as a test dynamical pattern. Consider again a set of single-variable training dynamical patterns $\varphi_{\zeta_y}^k$, $k = 1, \ldots, M$, with the kth training pattern $\varphi_{\zeta_y}^k$ generated from

$$\dot{x}_\zeta^k = F_\zeta^k \left(x_\zeta^k \right) \tag{7.30}$$

$$y_\zeta^k = x_{\zeta 1}^k \tag{7.31}$$

where $x_\zeta^k = [x_{\zeta 1}^k, \ldots, x_{\zeta n}^k]^T$ are the state variables of the kth training pattern $\varphi_{\zeta_y}^k$, y_ζ^k is the output variable that is available from measurement, $F^k(x_\zeta) = [x_{\zeta 2}^k, \ldots, x_{\zeta n}^k, f_\zeta^k(x_\zeta^k)]^T$ with $f_\zeta^k(x_\zeta^k)$ being an unknown smooth nonlinear function.

To achieve recognition of the test dynamical pattern from a set of training dynamical patterns, one possible method is to identify the system dynamics of the test dynamical pattern (as done for training dynamical patterns), and then compare the static, graph-based representations corresponding to the test and training dynamical patterns. It is known that to search for a match for the graph-based representations is the intractable isomorphism problem which is likely to be too computationally demanding for the time available [193].

The problem formulation is: without identifying the system dynamics of the test pattern φ_{ς_y}, search *rapidly* from the training single-variable dynamical patterns $\varphi_{\zeta_y}^k$ ($k = 1, \ldots, M$) for those *similar* to the given test single-variable dynamical pattern φ_{ς_y} in the sense of Definition 7.2 or 7.3.

For rapid recognition of a test single-variable dynamical pattern from a set of training single-variable dynamical patterns, another method is to first observe the states of the test pattern using a high-gain observer, and then achieve rapid recognition as in Chapter 5. This method is simple and feasible. Here we propose an approach using non-high-gain state observation. Since the system dynamics $f_\zeta^k(x_\zeta^k)$ of a training dynamical pattern $\varphi_{\zeta_y}^k$ can be accurately identified and stored in constant RBF network $\overline{W}^{k^T} S(Z)$, we construct a set of RBFN-based nonlinear observers as follows:

$$\dot{\hat{x}}_1^k = \hat{x}_2^k + k_1(y_\varsigma - \hat{y}^k)$$
$$\dot{\hat{x}}_2^k = \hat{x}_3^k + k_2(y_\varsigma - \hat{y}^k)$$
$$\vdots \qquad\qquad (7.32)$$
$$\dot{\hat{x}}_{n-1}^k = \hat{x}_n^k + k_{n-1}(y_\varsigma - \hat{y}^k)$$
$$\dot{\hat{x}}_n^k = \overline{W^k}^T S(\hat{x}^k) + k_n(y_\varsigma - \hat{y}^k)$$
$$\hat{y}^k = \hat{x}_1^k$$

where $k = 1, \ldots, M$, the superscript $(\cdot)^k$ denotes the component for the kth training pattern, $K = [k_1, \ldots, k_n]^T$ are observer gains, $\hat{x}^k = [\hat{x}_1^k, \ldots, \hat{x}_n^k]^T$ are estimates of the state x_ς, $\hat{y}^k$ denotes the estimate of system output y_ς of the test pattern, and the constant RBF network $\overline{W^k}^T S(\hat{x}^k)$ is embedded to provide a locally accurate approximation of system dynamics $f^k(x^k)$ of the training dynamical pattern $\varphi_{\zeta_y}^k$. These observers are taken as *dynamic representations* for the corresponding training single-variable dynamical patterns.

When a test single-variable dynamical pattern φ_{ς_y} is presented to one RBFN-based observer (i.e., the dynamical model for training pattern $\varphi_{\zeta_y}^k$), a state observation error system (i.e., recognition error system) is yielded as follows:

$$\dot{e}^k = (A - KC^T)e^k + B(f_\varsigma(x_\varsigma) - \overline{W^k}^T S(\hat{x}^k)) \qquad (7.33)$$

where $e^k = x_\varsigma - \hat{x}^k$ is the state estimation error, and (A, B, C) are the same as in (7.22).

REMARK 7.4

Note that in Section 7.3, the difference in the error system (7.22) contains the system dynamics $f(x)$ and its approximation $\overline{W}^T S(\hat{x})$. In the above state observation error system (7.33), the difference is expressed in terms of the system dynamics $f_\varsigma(x_\varsigma)$ of the test dynamical pattern φ_{ς_y} and the approximated dynamics $\overline{W^k}^T S(\hat{x}^k)$ of one training dynamical pattern. This implies that the analysis of stability and convergence of the above error system (7.33) will be more involved.

The problem formulation now becomes: among the set of RBFN-based observers (7.32), find the one that yields the smallest observation error. The corresponding training single-variable dynamical pattern $\varphi_{\zeta_y}^k$ will be considered as most similar to the test single-variable dynamical pattern φ_{ς_y}.

Without identifying the system dynamics of the test pattern φ_{ς_y}, the difference of system dynamics between the test and training patterns is not available from computation. Nevertheless, by conducting stability and convergence analysis for the recognition error system (7.33) using results from the nonlinear Lipschitz observer [188], it is proven that the state observation errors $\|e^k\|$

$(k = 1, \ldots, M)$ are approximately proportional to the differences of system dynamics between the test and training dynamical patterns. This difference can be explicitly measured by the state observation error $\|e^k\|$.

Using the result of non-high-gain state observation in Section 7.3, the following theorem describes how to achieve rapid recognition of a test single-variable dynamical pattern.

THEOREM 7.3

Consider the recognition error system (7.33) corresponding to the test pattern φ_{ς_y} and the dynamical model (RBFN observer) for the training pattern $\varphi^k_{\zeta_y}$. If the observer gain K is chosen such that the matrix $(A - KC^T)$ is stable and all the eigenvalues λ of $(A - KC^T)$ satisfy (7.23), and so the estimated state $\hat{x}^k$ stays with a local region Ω_ς along the orbit of the test pattern φ_{ς_y}, then the observation error $\|e^k\|$ will be approximately proportional to the difference between system dynamics of test pattern φ_{ς_y} and training pattern $\varphi^k_{\zeta_y}$.

PROOF By conducting stability and convergence analysis of the RBFN-based observer using nonlinear Lipschitz observer design [188], the problem of rapid recognition of the test single-variable dynamical pattern is turned into a problem of non-high-gain state observation, that is, to observe the full states of the test pattern φ_{ς_y} by using the set of RBFN-based observers, where the observer gains K are kept the same for $k = 1, \ldots, M$.

Note that when K is chosen as high gain, the estimated state $\hat{x}^k$ of the RBFN-based observers will converge closely to the state of the test pattern φ_{ς_y}. If the observer gain K is not so high but appropriately chosen, the estimated state $\hat{x}^k$ will stay with a local region Ω_ς along the orbit of the test pattern φ_{ς_y}; that is, $\text{dist}(\hat{x}^k, \varphi_{\varsigma_y}) < d_y$ where d_y is a positive constant.

From Equations (7.27) and (7.32), we have

$$\dot{e}^k = (A - KC^T)e^k + B[f_\varsigma(x_\varsigma) - f_\varsigma(\hat{x}^k)] + B[f_\varsigma(\hat{x}^k) - \overline{W}^{k^T} S(\hat{x}^k)] \quad (7.34)$$

For the error dynamics (7.34), consider Lyapunov function $V^k = e^{k^T} P e^k$. Its derivative satisfies

$$\dot{V}^k = e^{k^T}[(A - KC^T)^T P + P(A - KC^T)]e^k + 2e^{k^T} P[B(f_\varsigma(x_\varsigma) - f_\varsigma(\hat{x}^k))]$$
$$+ 2e^{k^T} P[B(f_\varsigma(\hat{x}^k) - \overline{W}^{k^T} S(\hat{x}^k))]$$

Since $x_\varsigma(t)$ is bounded, the stability and convergence analysis of the RBFN-based observer (7.34) can be conducted by borrowing the results of nonlinear Lipschitz observers [188]. Specifically, if the observer gain K is chosen such that the matrix $(A - KC^T)$ is stable and all the eigenvalues λ of $(A - KC^T)$ satisfy Equation (7.23), then, according to [188, Theorem 5], inequality (7.24) holds. According to [188, Theorem 3], there exists a symmetric positive definite matrix P and a constant $\varpi > 0$ such that

$$(A - KC^T)^T P + P(A - KC^T) + \gamma^2 PP + I + \varpi I = 0 \quad (7.35)$$

Then, we have

$$\dot{V}^k \leq e^{k^T}[(A - KC^T)^T P + P(A - KC^T) + \gamma^2 PP + I]e^k$$
$$+ 2e^{k^T} P[B(f_\varsigma(\hat{x}^k) - \overline{W}^{k^T} S(\hat{x}^k))]$$
$$\leq -\varpi \|e^k\|^2 + 2\|e^k\|\|PB\|\,|f_\varsigma(\hat{x}^k) - \overline{W}^{k^T} S(\hat{x}^k)|$$
$$\leq -\frac{1}{2}\varpi \|e^k\|^2 + \frac{2(\|PB\|\,|f_\varsigma(\hat{x}^k) - \overline{W}^{k^T} S(\hat{x}^k)|)^2}{\varpi}$$
$$\leq -\frac{\varpi}{2\lambda_{\max}(P)}\lambda_{\max}(P)\|e^k\|^2 + \frac{2\|PB\|(\varepsilon_y^{k^*} + \xi_y^{k^*})^2}{\varpi}$$
$$\leq -\frac{\varpi}{2\lambda_{\max}(P)}V^k + \frac{2\|PB\|^2(\varepsilon_y^{k^*} + \xi_y^{k^*})^2}{\varpi}$$
$$\leq -\alpha V^k + \delta \qquad (7.36)$$

where

$$\alpha := \frac{\varpi}{2\lambda_{\max}(P)}$$

$$\delta := \frac{2\|PB\|^2(\varepsilon_y^{k^*} + \xi_y^{k^*})^2}{\varpi}$$

$$\rho := \frac{\delta}{\alpha} = \lambda_{\max}(P)\frac{4\|PB\|^2(\varepsilon_y^{k^*} + \xi_y^{k^*})^2}{\varpi^2}$$

Then, Equation (7.36) gives

$$\lambda_{\min}(P)\|e^k\|^2 \leq V^k(t) < \rho + (V^k(0) - \rho)\exp(-\alpha t) \qquad (7.37)$$

That is,

$$\lambda_{\min}(P)\|e^k\|^2 < \rho + (V_z(0) - \rho)\exp(-\alpha t)$$
$$< \rho + V_z(0)\exp(-\alpha t) \qquad (7.38)$$

and

$$\|e^k\|^2 < (\rho + V_z(0)\exp(-\alpha t))/\lambda_{\min}(P) \qquad (7.39)$$

which implies that given $v > \sqrt{\frac{\lambda_{\max}(P)}{\lambda_{\min}(P)}} \cdot \frac{z\|PB\|(\varepsilon_y^* + \xi_y^*)}{\varpi}$, there exists a finite time T, such that for all $t \geq T$, the state observation error $\|e^k\|$ will converge exponentially to a neighborhood of zero; that is, $\|e^k\| \leq v$, with the size of the neighborhood v approximately proportional to $\varepsilon_y^{k^*} + \xi_y^{k^*}$, and inversely proportional to ϖ. Thus, the state observation error $\|e^k\|$ will be approximately proportional to the difference between the system dynamics of test pattern φ_{ς_y} and training pattern $\varphi_{\varsigma_y}^k$. ∎

REMARK 7.5
Note that the state variables $x_{\varsigma 2}, \ldots, x_{\varsigma n}$ of the system (7.27) generating the test pattern φ_{ς_y} are not measurable, so they cannot be used to compute the observation errors $\|e^k\|$. To solve this problem, a high-gain observer needs to be employed again to provide an accurate estimate of these state variables, so that the observation errors $\|e^k\|$ can be obtained. To achieve recognition using completely non-high-gain observation, a possible method is to make a decision based only on $|e_1^k(t)|$. A detailed analysis of the method requires more study.

From the above analysis, it is seen that the difference between system dynamics of the test and training single-variable dynamical patterns can be explicitly measured by $\|e^k\|_{t \geq T_0}$ (for short T_0). Thus, we take the following method to rapidly recognize a test single-variable dynamical pattern from a set of training single-variable dynamical patterns:

1. Identify the system dynamics of a set of training single-variable dynamical patterns $\varphi_{\varsigma_y}^k \ k = 1, \ldots, M$ using deterministic learning and high-gain observation.

2. Construct a set of RBFN-based observers (7.32) as dynamic representations for the training single-variable dynamical patterns $\varphi_{\varsigma_y}^k$.

3. Take the state y_ς of a test single-variable pattern φ_{ς_y} as the RBFN input to the dynamical models (7.32), and compute the average l_1 norm of the state observation error $\|e^k(t)\|$

$$\|e^k\|_{t_1} = \frac{1}{t} \int_{t_0}^{t_0+t} \|e^k\| dt, \qquad i = 1, \ldots, n \qquad (7.40)$$

4. Take the training single-variable dynamical pattern whose corresponding RBFN observer yields the smallest $\|e^k\|_{t_1}$ as the one most *similar* to the test single-variable dynamical pattern φ_{ς_y} in the sense of Definition 7.3.

7.5   Simulation Studies

Consider the well-known van der Pol oscillator [28,227] already considered in Chapters 4 and 6:

$$\dot{x}_1 = x_2$$
$$\dot{x}_2 = -x_1 + \beta\left(1 - x_1^2\right)x_2$$
$$y = x_1 \qquad (7.41)$$

where $x = [x_1, x_2]^T$ is the state, β is a constant parameter, and the system dynamics $f(x) = -x_1 + \beta(1 - x_1^2)x_2$ is an unknown, smooth nonlinear function. The van der Pol oscillator is presented in the form of system (7.27) by choosing x_1 to be the output. For initial states starting from points other than $[0, 0]$, the van der Pol oscillator can yield a limit cycle trajectory when $\beta > 0$.

Identification and representation: A *single-variable dynamical pattern* is the periodic or periodic-like (recurrent) system output trajectory $y(t)$. We consider two single-variable dynamical patterns generated from system (7.41), as shown in Figures 7.1a and b. Denoted as $\varphi^1_{\zeta_y}$ and $\varphi^2_{\zeta_y}$, the two single-variable periodic dynamical patterns are started from initial states $x(0) = [x_1(0), x_2(0)]^T = [0.5, -1.0]^T$, with system parameters $\beta = 0.2$ and $\beta = 0.8$, respectively.

By using the high-gain observer (7.2) ($n = 2$), with design parameters chosen as $h_1 = h_2 = 1$ and $k = 40$, accurate state observation of the state x_2 of the van der Pol oscillator is achieved for both dynamical patterns $\varphi^1_{\zeta_y}$ and $\varphi^2_{\zeta_y}$, as shown in Figures 7.1c and d. The observation errors are shown in Figures 7.1e and f.

To identify the unknown dynamics $f(x)$ of the two patterns $\varphi^1_{\zeta_y}$ and $\varphi^2_{\zeta_y}$, the dynamical RBF network (7.3) is employed. The RBF network $\widehat{W}^T S(x)$ is constructed in a regular lattice, with nodes $N = 441$, the centers μ_i evenly spaced on $[-3.0, 3.0] \times [-3.0, 3.0]$, and the widths $\eta_i = 0.3$. The weights of the RBF networks are updated according to Equation (7.4). The design parameters for Equations (7.3) and (7.4) are $a = 5$, $\Gamma = 2$, and $\sigma = 0.001$. The initial weights $\widehat{W}(0) = 0$.

The phase portrait of dynamical pattern φ^1_ζ is shown in Figure 7.2a. Its corresponding system dynamics $f(x)$ is shown in Figure 7.2b. In Figure 7.2c, it is seen that some weight estimates (of the neurons whose centers are close to the orbit of the pattern) converge to constant values, whereas some other weight estimates (of neurons centered far away from the orbit) remain almost zero. The locally accurate NN approximation of $f(x; p)$ along the orbit of the periodic pattern φ^1_ζ is clearly shown in Figures 7.2d and e. In Figure 7.2f, dynamical pattern $\varphi^1_{\zeta_y}$ is represented by the constant RBF network $\overline{W}^T S(Z)$. This representation is time-invariant, based on the fundamental information of the system dynamics. It is also spatially distributed, involving a large number of neurons distributed along the orbit of the dynamical pattern. The NN approximation is accurate only in the vicinity of the periodic pattern. Away from this region, where the orbit of the pattern does not explore, no learning occurs, as shown by the zero-plane in Figure 7.2f, that is, the small values of $\overline{W}^T S(Z)$ in the unexplored area.

Similarly, from Figures 7.3a and b, we can see the phase portrait and the system dynamics $f(x)$ of pattern $\varphi^2_{\zeta_y}$. Figure 7.3c shows the partial parameter convergence. The locally accurate NN approximation of system dynamics $f(x)$ along the orbit of the pattern is shown in Figures 7.3d and e. Figure 7.3f shows the time-invariant representation of pattern $\varphi^2_{\zeta_y}$.

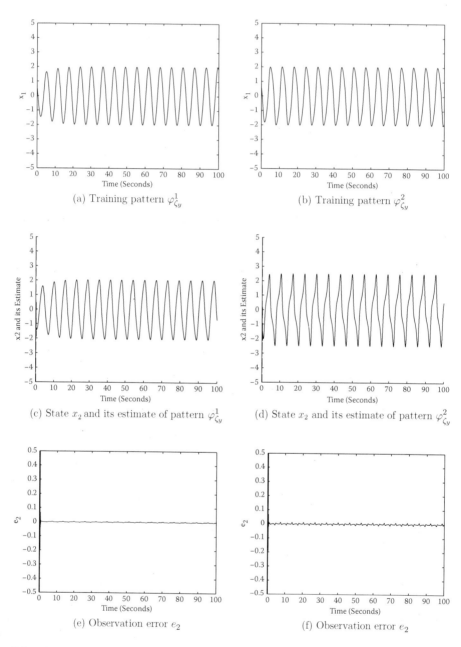

(a) Training pattern $\varphi_{\zeta_y}^1$

(b) Training pattern $\varphi_{\zeta_y}^2$

(c) State x_2 and its estimate of pattern $\varphi_{\zeta_y}^1$

(d) State x_2 and its estimate of pattern $\varphi_{\zeta_y}^2$

(e) Observation error e_2

(f) Observation error e_2

FIGURE 7.1
High-gain observation of single-variable dynamical patterns.

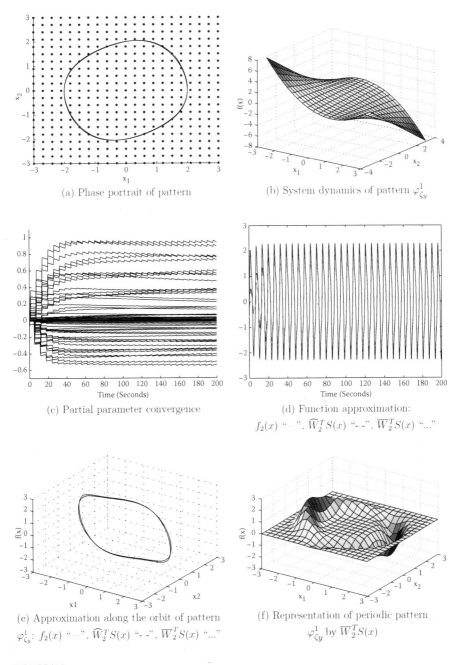

(a) Phase portrait of pattern

(b) System dynamics of pattern $\varphi^1_{\zeta_y}$

(c) Partial parameter convergence

(d) Function approximation:
$f_2(x)$ "—", $\widehat{W}_2^T S(x)$ "- -", $\overline{W}_2^T S(x)$ "..."

(e) Approximation along the orbit of pattern
$\varphi^1_{\zeta_y}$: $f_2(x)$ "—", $\widehat{W}_2^T S(x)$ "- -", $\overline{W}_2^T S(x)$ "..."

(f) Representation of periodic pattern
$\varphi^1_{\zeta_y}$ by $\overline{W}_2^T S(x)$

FIGURE 7.2
Deterministic learning of training pattern $\varphi^1_{\zeta_y}$.

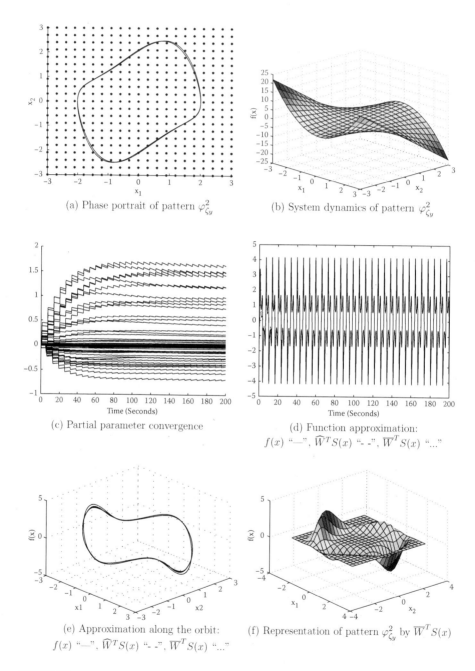

(a) Phase portrait of pattern $\varphi_{\zeta y}^2$

(b) System dynamics of pattern $\varphi_{\zeta y}^2$

(c) Partial parameter convergence

(d) Function approximation:
$f(x)$ "—", $\widehat{W}^T S(x)$ "- -", $\overline{W}^T S(x)$ "..."

(e) Approximation along the orbit:
$f(x)$ "—", $\widehat{W}^T S(x)$ "- -", $\overline{W}^T S(x)$ "..."

(f) Representation of pattern $\varphi_{\zeta y}^2$ by $\overline{W}^T S(x)$

FIGURE 7.3

Deterministic learning of chaotic pattern $\varphi_{\zeta y}^2$.

Rapid recognition: two periodic patterns (as shown in Figure 7.4) are used as the test single-variable dynamical patterns and are denoted as $\varphi_{\zeta_y}^1$ and $\varphi_{\zeta_y}^2$, respectively. Test pattern $\varphi_{\zeta_y}^1$ is generated from system (7.41), with initial states $x(0) = [x_1(0), x_2(0)]^T = [0.0, -1.8]^T$, and system parameters $\beta = 0.1$. The test pattern $\varphi_{\zeta_y}^2$ is generated from system (7.41), with initial states $x(0) = [x_1(0), x_2(0)]^T = [0.1, -1.7]^T$, and system parameters $\beta = 0.7$.

As single-variable dynamical patterns $\varphi_{\zeta_y}^1$ and $\varphi_{\zeta_y}^2$ have been locally accurately identified via high-gain observation and deterministic learning, they are taken as two training dynamical patterns. Two RBFN-based non-linear observers (observers 1 and 2) are constructed according to (7.32) as dynamical representatives for the two training patterns. The time-invariant representations $\overline{W}^{k^T} S(Z)$ ($k = 1, 2$) obtained above are embedded into the two RBFN-based nonlinear observers (7.32). The observer gains are chosen as $k_1 = 2$, $k_2 = 20$, which are much smaller than $h_1 k = 40$, $h_2 k^2 = 1600$ used above.

First, consider the recognition of test pattern $\varphi_{\zeta_y}^1$ by training patterns $\varphi_{\zeta_y}^1$ and $\varphi_{\zeta_y}^2$. Figures 7.5a and b show the system dynamics $f(x; p) = -x_1 + \beta(1 - x_1^2)x_2$ along the estimated orbit of test pattern $\varphi_{\zeta_y}^1$, together with the RBFN approximations of the system dynamics of the training patterns $\varphi_{\zeta_y}^1$ and $\varphi_{\zeta_y}^2$, respectively. The observation errors $\|e^k(t)\|$ ($k = 1, 2$) are shown in Figures 7.5c and d. The average l_1 norms of the observation errors, that is, $\|e^k(t)\|_{l_1}$ ($k = 1, 2$), are shown in Figures 7.5e and 7.5f. It is clearly seen in Figure 7.5f that from the beginning stage of the recognition process, $\|e^1(t)\|_{l_1}$ is smaller than $\|e^2(t)\|_{l_1}$. Because the same observer gains are used for observers 1 and 2, it is concluded that test pattern $\varphi_{\zeta_y}^1$ is more similar to training pattern $\varphi_{\zeta_y}^1$ than to training pattern $\varphi_{\zeta_y}^2$.

Similarly, in recognition of test dynamical pattern $\varphi_{\zeta_y}^2$, Figures 7.6a and b show the system dynamics $f(x; p) = -x_1 + \beta(1 - x_1^2)x_2$ along the estimated orbit of test pattern $\varphi_{\zeta_y}^2$, together with the RBFN approximations of the system dynamics of the training patterns $\varphi_{\zeta_y}^1$ and $\varphi_{\zeta_y}^2$, respectively. The observation errors $\|e^k(t)\|$ ($k = 1, 2$) are shown in Figures 7.6c and d. It is seen in Figure 7.6f that from the beginning stage of the recognition process, $\|e^2(t)\|_{l_1}$ is smaller than $\|e^1(t)\|_{l_1}$. Thus, test pattern $\varphi_{\zeta_y}^2$ is recognized as more similar to training pattern $\varphi_{\zeta_y}^2$ than to training pattern $\varphi_{\zeta_y}^1$.

From Figures 7.5f to 7.6f, it is also seen that comparison of state observation can be achieved within a very short period of time, which means that the test single-variable patterns are rapidly recognized as similar or dissimilar to training single-variable patterns.

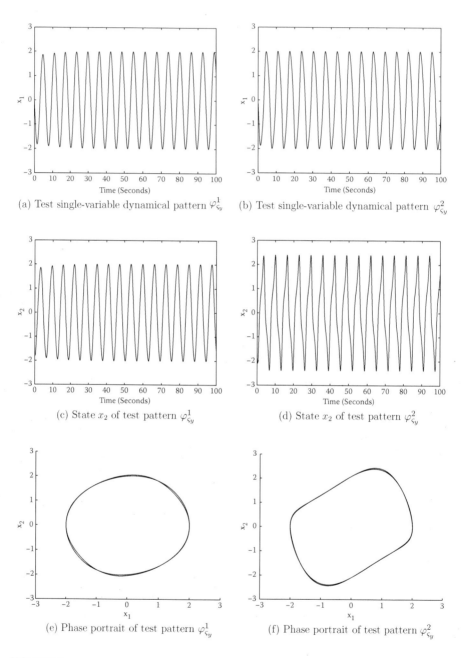

(a) Test single-variable dynamical pattern $\varphi^1_{\varsigma_y}$

(b) Test single-variable dynamical pattern $\varphi^2_{\varsigma_y}$

(c) State x_2 of test pattern $\varphi^1_{\varsigma_y}$

(d) State x_2 of test pattern $\varphi^2_{\varsigma_y}$

(e) Phase portrait of test pattern $\varphi^1_{\varsigma_y}$

(f) Phase portrait of test pattern $\varphi^2_{\varsigma_y}$

FIGURE 7.4

Test single-variable dynamical patterns $\varphi^1_{\varsigma_y}$ and $\varphi^2_{\varsigma_y}$.

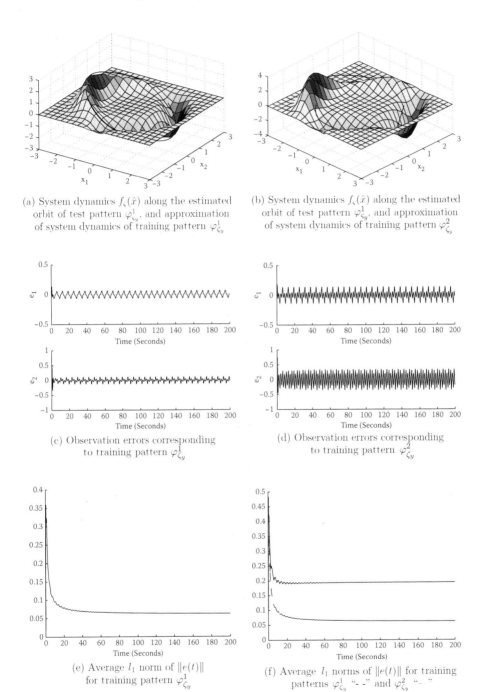

(a) System dynamics $f_\varsigma(\hat{x})$ along the estimated orbit of test pattern $\varphi^1_{\zeta_y}$, and approximation of system dynamics of training pattern $\varphi^1_{\zeta_y}$

(b) System dynamics $f_\varsigma(\hat{x})$ along the estimated orbit of test pattern $\varphi^1_{\zeta_y}$, and approximation of system dynamics of training pattern $\varphi^2_{\zeta_y}$

(c) Observation errors corresponding to training pattern $\varphi^1_{\zeta_y}$

(d) Observation errors corresponding to training pattern $\varphi^2_{\zeta_y}$

(e) Average l_1 norm of $\|e(t)\|$ for training pattern $\varphi^1_{\zeta_y}$

(f) Average l_1 norms of $\|e(t)\|$ for training patterns $\varphi^1_{\zeta_y}$ "- -" and $\varphi^2_{\zeta_y}$ "- "

FIGURE 7.5
Recognition of test pattern $\varphi^1_{\zeta_y}$ by training patterns $\varphi^1_{\zeta_y}$ and $\varphi^2_{\zeta_y}$.

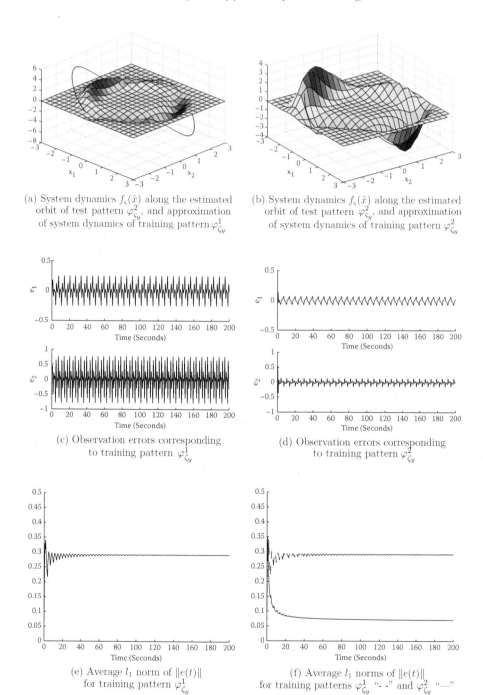

(a) System dynamics $f_{\varsigma}(\hat{x})$ along the estimated orbit of test pattern $\varphi^2_{\varsigma_y}$, and approximation of system dynamics of training pattern $\varphi^1_{\varsigma_y}$

(b) System dynamics $f_{\varsigma}(\hat{x})$ along the estimated orbit of test pattern $\varphi^2_{\varsigma_y}$, and approximation of system dynamics of training pattern $\varphi^2_{\varsigma_y}$

(c) Observation errors corresponding to training pattern $\varphi^1_{\varsigma_y}$

(d) Observation errors corresponding to training pattern $\varphi^2_{\varsigma_y}$

(e) Average l_1 norm of $\|e(t)\|$ for training pattern $\varphi^1_{\varsigma_y}$

(f) Average l_1 norms of $\|e(t)\|$ for training patterns $\varphi^1_{\varsigma_y}$ "- -" and $\varphi^2_{\varsigma_y}$ "—"

FIGURE 7.6
Recognition of test pattern $\varphi^2_{\varsigma_y}$ by training patterns $\varphi^1_{\varsigma_y}$ and $\varphi^2_{\varsigma_y}$.

7.6 Summary

In this chapter, we have shown that the deterministic learning mechanism can be utilized to improve nonlinear observer design in the sense of allowing both accurate state estimation and system identification. For a class of nonlinear systems undergoing periodic or recurrent motions with only output measurements, first, by using a high-gain observer and the deterministic learning mechanism, locally accurate identification of system dynamics has been achieved along the estimated system states. Second, the learned knowledge of system dynamics has been reused in an RBFN-based nonlinear observer to achieve non-high-gain design. In this way, the difficult problem of nonlinear observer design can be successfully resolved by incorporating the deterministic learning mechanisms.

The improved nonlinear observer technique can be further used in other related areas such as dynamic fault diagnosis and dynamical pattern recognition. By learning the underlying system dynamics of a set of training dynamical patterns first, and then constructing a set of nonlinear observers as representatives of the training patterns, rapid recognition of a test dynamical pattern has been implemented. Moreover, the recognition is achieved not by using high gains, but according to a kind of internal and dynamical matching of system dynamics. The observation errors are taken as the measure of similarity between the test and training dynamical patterns.

Note that the internal and dynamical matching of system dynamics is what we refer to in Chapter 5 as the dynamical parallel distributed processing (DPDP), which is also implemented in a continuous and analog manner. The non-high-gain observation makes the differences of system dynamics explicitly unfolded in time. The significance of this research lies in that an observation-based approach has been proposed for dynamical pattern processing in which the problem of rapid recognition of single-variable dynamical patterns is turned into the problem of non-high-gain state observation.

8

Toward Human-Like Learning and Control

In the preceding chapters, it has been shown that the proposed deterministic learning (DL) theory is closely related to many areas in the discipline of systems and control, such as system identification, adaptive control, intelligent control, and nonlinear observer design. It is developed using concepts and tools from these areas. The significance of determanistic learning lies in providing a unified conceptual framework for knowledge acquisition, representation, and utilization in uncertain dynamic environments. Moreover, improved understanding of the employed concepts (e.g., persistence of excitation [PE]) in systems and control has occurred, and approaches of systematic design for system identification, pattern recognition, and intelligent control of nonlinear systems have been suggested, which potentially advance substantially the above-mentioned systems and control areas.

Of particular interest is that the overall framework has many characteristics of human-like learning and control capabilities. The further development needed can usefully explore this aspect. With this in mind, this final chapter draws conclusions and makes suggestions for further work.

8.1 Knowledge Acquisition

First, deterministic learning theory implements knowledge acquisition in processes of nonlinear system identification, closed-loop NN control of nonlinear systems, and state observation of nonlinear systems. Key elements to achieve knowledge acquisition include: (i) employment of the localized radial basis function network (RBFN), (ii) satisfaction of a partial PE condition along a periodic or recurrent orbit, (iii) guaranteed exponential stability of the linear time-varying (LTV) adaptive systems, and (iv) accurate RBFN approximation of unknown nonlinear dynamics achieved in a local region along the recurrent orbit.

In conventional system identification, the convergence to true parameters and the identification of the corresponding system model relies on the satisfaction of the PE condition. However, it was found that while for linear system identification the PE condition can be satisfied when the input signal is sufficiently rich in the frequency domain, there is no general relationship established between the frequencies of the input signals and the parameters to be estimated for nonlinear system identification. Consequently, identification for a true nonlinear system model is very difficult to be achieved. Closed-loop identification is then studied for the purpose of model-based control, in which the acceptance of the identified models is justified by the "usefulness" rather than "truth." In other words, identification of true closed-loop system models is also a very difficult problem.

In DL-based identification for nonlinear systems, the difficulty of identifying the true system model is handled by selecting localized RBF networks as the parameterized model structure. When a recurrent orbit is taken as the input to the RBF network, a direct connection between the recurrent NN input to the estimated weights of neurons centered in a local region along the periodic or periodic-like orbit is established. This leads naturally to the satisfaction of a partial PE condition and subsequently exponential stability of the LTV adaptive systems. Consequently, partial parameter convergence and locally accurate identification of a partial true system model are achieved in a local region along the periodic or periodic-like orbit.

In DL-based NN control of nonlinear systems, it has been shown that an appropriately designed adaptive NN controller is capable of identifying closed-loop system dynamics during tracking control to a periodic or periodic-like reference orbit. Accurate NN approximation for closed-loop system dynamics can be achieved in a local region along the periodic or periodic-like state trajectory. Therefore, even for closed-loop identification for model-based control, the partial true closed-loop system model can be locally accurately identified via deterministic learning. Furthermore, for identification and control of nonlinear systems with output measurements, by combining deterministic learning with a nonlinear high-gain observer technique, the estimated state information can be used to accurately identify the underlying system dynamics in a nonlinear observer problem. Accurate identification of system dynamics is achieved in a local region along the estimated state trajectory.

In summary, DL theory is capable of obtaining fundamental knowledge about system dynamics from uncertain dynamic processes. The nature of knowledge acquisition is related to the exponential stability of a certain class of LTV adaptive systems, which is ensured by the satisfaction of a partial PE condition. The learned knowledge about system dynamics can be stored and represented by constant RBF networks, and can be reused to implement rapid recognition of dynamical patterns, or to achieve improved control performance.

8.2 Representation and Similarity

In dynamical processes such as dynamical pattern recognition and feedback control, one important issue is how to appropriately represent the time-varying patterns or control situations. This issue becomes more difficult when the representation is to be presented in a time-independent manner. Instead of using a limited number of features extracted from measurements or observations as in static pattern recognition, a dynamical pattern or control situation can be effectively represented in a time-invariant and spatially distributed manner using the knowledge obtained from deterministic learning. Complete information on both the pattern state and the underlying system dynamics is utilized for representation of a dynamical pattern. The time-invariant representation is a kind of static representation stored in constant RBF networks. Using these constant RBF networks, a set of nonlinear dynamic models is constructed as dynamic representations of the training dynamical patterns.

Another important issue in dynamical environments is the characterization of similarity between two dynamical patterns or control situations. The existing similarity measures based on distances for static patterns might be inappropriate to define the similarity of dynamical patterns. We propose a similarity definition based on the qualitative analysis of nonlinear dynamical systems. Specifically, similarity of two dynamical patterns or control situations is characterized based on the difference between the system dynamics inherently within the dynamical patterns. This definition is in accordance with the concepts of *topological equivalence* and *structural stability* in dynamical system theory.

8.3 Knowledge Utilization

Deterministic learning consists of the phases of knowledge acquisition and knowledge utilization. The value of the acquired knowledge can be manifested only through utilization of the knowledge in dynamic processes, for example, rapid recognition of dynamical patterns, pattern-based learning control, and non-high-gain state observation.

In rapid recognition of a test dynamical pattern from a set of training dynamical patterns, use is made of the knowledge of system dynamics of training dynamical patterns being represented in the form of constant RBF networks. The constant RBF networks are then embedded into a set of state estimators. For a test dynamical pattern, if its underlying system dynamics is topologically similar to that of one training dynamical pattern, state estimation or synchronization will be achieved according to a kind of internal and dynamical matching on system dynamics, and the test pattern is recognized as similar to the training dynamical pattern. The estimation or synchronization errors can be taken as the measure of similarity between the test and training patterns.

In pattern-based learning control, a pattern-based NN controller can effectively recall and reuse the learned knowledge to conduct an internal and dynamical matching of system dynamics underlying similar control situations. Stability and improved control performance can be achieved without readapting to the uncertainties in the closed-loop control process. In nonlinear observer design, the learned knowledge on system dynamics can be reused so that correct state estimation can be achieved without using high gains. Moreover, the improved nonlinear observer technique can be applied to resolve the problem of rapid recognition of single-variable dynamical patterns via non-high-gain state observation achieved again according to dynamical matching on system dynamics.

It is seen that the previously learned knowledge can be utilized to compare the similarity of dynamical patterns or control situations via the so-called internal and dynamical matching of system dynamics. This actually represents a new model of information processing, which we refer to as dynamical parallel distributed processing (DPDP). It is seen that the learned knowledge is utilized in completely dynamical processes. The nature of the knowledge utilization in dynamic environments is related to the stability and convergence of certain classes of perturbed linear time-invariant (LTI) systems.

8.4   Toward Human-Like Learning and Control

Deterministic learning theory provides a unified approach to human-like learning and control. Humans are generally good at temporal/dynamical pattern recognition in that the information distributed over time underlying dynamical patterns can be effectively identified, represented, recognized, and classified. The recognition process takes place quickly from the beginning of sensing temporal patterns, and runs directly on the input space for feature extraction and pattern matching. Humans can also learn many highly complicated control tasks with sufficient practice, and perform these tasks again and again with little effort. Experiments demonstrate that humans learn the dynamics of reaching movements through a flexible combination of primitives that have Gaussian-like tuning functions [222]. Moreover, the motor control system builds a model (called the internal model) of the environment as a map between the experienced somatosensory input and the output forces needed to counterbalance the external perturbations. In addition, results indicate that this internal model is valid locally near the experienced motion trajectory; it smoothly decays with distance from the perturbed locations [60]. These human learning and control mechanisms, although not fully understood, appear to be quite different from the conventional approaches in the literature for learning and control.

With the development of deterministic learning theory, the pattern-based learning and control framework appears to be consistent with mechanisms of human learning and control. In the process of tracking control to a recurrent reference trajectory, an appropriately designed adaptive NN controller using Gaussian RBF networks can develop internal models of the external force fields. The learned internal models are locally accurate along the recurrent trajectory. They can be stored as knowledge and recalled to compute the required torques for similar control tasks. The DL-based framework bears similarity to proficient human control with little cognitive effort. It would be useful to explore further in areas such as motion learning and control of humanoid robotics.

8.5 Cognition and Computation

Deterministic learning theory may even provide insight to natural cognitive systems from the perspective of dynamics. Cognition and computation have been deeply linked for at least fifty years. The origin of the electronic digital computer lies in Turing's attempt to formalize the kinds of symbolic logical manipulations that human mathematicians can perform. Digital computation was later viewed as the correct conceptual framework for understanding cognition in general [168]. Another tradition for understanding cognition is rooted in dynamical systems theory. Dynamical approaches to cognition go back to the cybernetics era in the 1940s. Information theory, dynamics, and computation were brought together in studying the brain. Ashby made the startling proposal that all of cognition might be accounted for with dynamical system models [8]. However, with the dominance of symbolic AI in the 1960s and 1970s, dynamical systems-based approaches were not extensively pursued.

Recently, many proponents of dynamical approaches argue that computation is a misleading notion to use in understanding cognition. Van Gelder and Port [228] seek to show that the "computational approach" ("cognitive operations are transformations from one static symbol structure to the next") is false, and propose the "dynamical hypothesis" ("cognition is best understood in the language of dynamical systems theory"). However, little work directly followed from the speculation due to a lack of appropriate mathematical methods and tools to implement practical models.

Deterministic learning theory provides strong support to the dynamical systems hypotheses in cognitive science. Identification and recognition of dynamical patterns are indeed best understood from a viewpoint of stability analysis of LTV or LTI systems. The dynamical versions of localized RBF networks can be considered as reasonable models for natural cognitive systems due to their capabilities of knowledge acquisition, representation, and utilization in dynamic environments. Furthermore, the new model for information processing, that is, dynamical parallel distributed processing will probably lead to a renewed era of analog computation.

8.6   Comparison with Statistical Learning

Over the past decade, statistical learning has become the mainstream in the area of machine learning. Many problems in learning of static nonlinear mappings have been successfully resolved via statistical learning. For example, research on pattern recognition and even neural networks has been mainly conducted via the statistical approach [19,95,254]. In statistical learning, the learning problem is considered as function estimation on the basis of empirical data. The nature of statistical learning is revealed by considering the problem of estimating the values of an unknown function at given points of interest. Originally, this problem was attacked by first estimating the entire function at all points of the domain and second, estimating the function at the given points. It is obvious that one may not have enough information to estimate the function at all points. The physiology of statistical learning is then revealed by the goal "NOT to solve the problem of estimating the values of a function at given points by estimating the entire function" [229]. This physiology is related to the essence of human intelligence.

The deterministic learning theory is not developed using statistical principles. Assumptions on probability distributions are not necessary. Nonetheless, DL theory has some physiological similarities to statistical learning, in the sense that instead of achieving identification of a system model in the entire state space, accurate identification of a partial system model is achieved only in local regions. For the space where the recurrent orbit does not explore, no learning occurs, as represented by the slightly updated neural weights for neurons far away from the orbit and the small values of RBFN approximation in the unexplored area. This can be compared with not estimating all points of the nonlinear function in statistical learning and so coincides with the physiology of statistical learning in nature.

8.7   Applications of the Deterministic Learning Theory

The content of this monograph is justified by the objective to collect and expand the basic ideas and results. It is clearly seen that there is much more research needed in this new area. It should be acknowledged that there are numerous directions for further theoretical work. Extensions of the basic results to nonrecurrent orbits (i.e., patterns or tasks), other approximation networks, and more general dynamical systems should be given priority.

The power of the deterministic learning methodology for resolving difficult problems as well as for opening new directions indicates that it has the potential to become a new direction in areas of machine learning, system identification/modeling, pattern recognition, intelligent control, cognitive science, fault diagnosis, and so on. For instance, in the literature of fault diagnosis, although the problem of fault detection has been extensively investigated [50,130,223,231,232], the fault isolation (classification) problem has received less attention [267]. There have not been many analytical results on fault isolation and prediction, especially in the case of uncertain nonlinear systems. The presented deterministic learning theory, especially the approach for identification and rapid recognition of dynamical patterns, provides a solution for the problem of rapid isolation of oscillation faults generated from uncertain nonlinear systems [30]. The result may be further applied to recognition and analysis of ECG/EEG signals, prediction of epileptic seizures, and security assessment and pattern-based control of power systems.

References

1. B. D. O. Anderson, "Exponential stability of linear equations arising in adaptive identification," *IEEE Transactions on Automatic Control*, Vol. 22, no. 2, pp. 83–88, 1977.
2. B. D. O. Anderson, J. B. Moore, and R. M. Hawkes, "Model approximation via prediction error identification," *Automatica*, Vol. 14, pp. 615–622, 1978.
3. B. D. O. Anderson, and R. M. Johnstone, "Adaptive systems and time varying plants," *Int. J. of Control*, Vol. 37, no. 2, pp. 367–377, 1983.
4. B. D. O. Anderson, "Adaptive systems, lack of persistency of excitation and bursting phenomena," *Automatica*, Vol. 21, pp. 247–258, 1985.
5. B. D. O. Anderson, R. R. Bitmead, C. R. Johnson, P. V. Kokotovic, R. L. Kosut, I. Mareels, L. Praly, and B. Riedle, *Stability of Adaptive Systems*, MIT Press, Cambridge, Massachusetts, 1986.
6. P. J. Antsaklis, "Guest Editor's Introduction," *IEEE Control Systems Magazine*, Vol. 15, no. 3, pp. 5–7, June 1995; Special Issue on Intelligence and Learning, *IEEE Control Systems Magazine*, P. J. Antsaklis (Ed.), Vol. 15, no. 3, pp. 5–80, June 1995.
7. T. M. Apostol, *Mathematical Analysis*, Addison-Wesley, Reading, Massachusetts, 1963.
8. R. Ashby, *Design for a Brain*, Chapman-Hall, London, 1952.
9. K. J. Astrom, and T. Bohlin, "Numerical identification of linear dynamic systems from normal operating records," in *Proc. IFAC Symposium on Self-Adaptive Systems*, Teddington, UK, 1965, pp. 96–111.
10. K. J. Astrom, and P. Eykhoff, "System identification: A survey," *Automatica*, Vol. 7, pp. 123–162, 1971.
11. K. J. Astrom, "Maximum likelihood and prediction error methods," *Automatica*, Vol. 16, pp. 551–574, 1980.
12. K. J. Astrom, "Theory and applications of adaptive control, a survey," *Automatica*, Vol. 19, no. 5, pp. 471–486, 1983.
13. K. J. Astrom, and B. Wittenmark, *Adaptive Control*, Addison-Wesley, Reading, Massachusetts, 1989.
14. P. Ball, *The Self-Made Tapestry: Pattern Formation in Nature*, Oxford University Press, New York, 1999.
15. R. Barron, "Self-organizing control," *Control Engineering*, February-March, 1968.
16. A. R. Barron, "Approximation and estimation bounds for artificial neural networks," *Proc. 4th Ann. Workshop on Computational Learning Theory*, pp. 243–249, 1991.
17. R. E. Bellman, *Adaptive Control Processes: A Guided Tour*, Princeton University Press, Princeton, New Jersey, 1961.
18. D. Bestle, and M. Zeitz, "Canonical form observer design for nonlinear time-variable systems," *Int. J. Control*, Vol. 38, no. 2, pp. 419–431, 1983.

19. C. Bishop, *Neural Networks for Pattern Recognition*, Oxford University Press, New York, 1995.

20. R. R. Bitmead, "Persistence of excitation conditions and the convergence of adaptive schemes," *IEEE Transactions on Information Theory*, Vol. 30, no. 3, pp. 183–191, 1984.

21. S. Boyd, and S. Sastry, "Necessary and sufficient conditions for parameter convergence in adaptive control," *Automatica*, Vol. 22, no. 6, pp. 629–639, 1986.

22. D. S. Broomhead, and D. Lowe, "Multivariable functional interpolation and adaptive networks," *Complex Systems*, Vol. 2, pp. 321–355, 1988.

23. M. D. Buhmann, *Radial Basis Functions*, Cambridge University Press, Cambridge, 2003.

24. C. I. Byrnes, and A. Isidori, "New results and examples in nonlinear feedback stabilization," *System & Control Letters*, Vol. 12, pp. 437–442, 1989.

25. C. S. Chang, "Online transient stability evaluation of interconnected power systems using pattern recognition strategy," *IEE Proceedings C*, Vol. 140, no. 2, pp. 115–122, 1993.

26. F. C. Chen, and C. C. Liu, "Adaptively controlling nonlinear continuous-time systems using multilayer neural networks," *IEEE Transactions on Automatic Control*, Vol. 39, no. 6, pp. 1306–1310, 1994.

27. F. C. Chen, and H. K. Khalil, "Adaptive control of a class of nonlinear discrete-time systems using neural networks," *IEEE Transactions Automatic Control*, Vol. 40, no. 5, pp. 791–801, 1995.

28. G. Chen, and X. Dong, *From Chaos to Order: Methodologies, Perspectives and Applications*, World Scientific, Singapore, 1998.

29. J. Chen, and R. J. Patton, *Robust Model-Based Fault Diagnosis for Dynamic Systems*, Kluwer, Boston, Massachusetts, 1999.

30. T. R. Chen, and C. Wang, "Deterministic learning and oscillation fault diagnosis," *Proceedings of the 7th World Congress on Intelligent Control and Automation*, Chongqing, China, June 25–27, 2008.

31. E. W. Cheney, and W. Light, *A Course in Approximation Theory*, Brooks/Cole, Publishing Company, Pacific Grove, California, 2000.

32. J. Y. Choi, and J. A. Farrell, "Adaptive observer backstepping control using neural networks," *IEEE Transactions on Neural Networks*, Vol. 12, no. 5, pp. 1103–1112, 2001.

33. G. Ciccarella, M. D. Mora, and A. Germani, "A Luenberger-like observer for nonlinear systems," *Int. J. Control*, Vol. 57, no. 3, pp. 537–556, 1993.

34. E. Covey, H. L. Hawkins, and R. F. Port (Eds.), *Neural Representation of Temporal Patterns*, Plenum Press, New York, 1995.

35. G. Cybenko, "Approximation by superpositions of a sigmoidal function," *Mathematics of Control, Signals and Systems*, Vol. 2, no. 4, pp. 303–314, 1989.

36. F. Deza, D. Bossanne, E. Busvelle, J. P. Gauthier, and D. Rakotopara, "Exponential observers for nonlinear systems," *IEEE Transactions on Automatic Control*, Vol. 38, no. 3, pp. 482–484, 1993.

37. X. Ding, P. Frank, and L. Guo, "Nonlinear observer design via an extended observer canonical form," *Systems & Control Letters*, Vol. 15, p. 313, 1990.

38. D. Dochain, "State and parameter estimation in chemical and biochemical processes: A tutorial," *Journal of Process Control*, Vol. 13, pp. 801–818, 2003.

39. P. DuChateau, *Advanced Calculus*, HarperPerennial, New York, 1992.

40. G. Duffing, "Erzwungene Schwingungen bei veränderlicher Eigenfrequenz und ihre technische Bedeutung," *Vieweg*, Braunschweig, 1918.

41. N. Dyn, "Interpolation and approximation by radial and related functions," in *Approximation Theory* VI: Vol. I, C. K. Chui, L. L. Schumaker, and J. D. Ward, Eds., Academic Press, New York, 1989.

42. J. L. Elman, "Finding structure in time," *Cognitive Science*, Vol. 14, pp. 179–211, 1990.

43. S. Fabri, and V. Kadi, "Dynamics structure NNs for stable adaptive control of nonlinear systems," *IEEE Trans. Neural Networks*, Vol. 7, no. 5, pp. 1151–1167, 1996.

44. J. Farrell, and W. Baker, "Learning control systems," in *Introduction to Intelligent and Autonomous Control*, K. M. Passino and P. J. Antsaklis, Eds., Kluwer Academic, Norwell, Massachusetts, 1993.

45. J. Farrell, "Persistence of excitation conditions in passive learning control," *Automatica*, Vol. 33, pp. 699–703, 1997.

46. J. Farrell, "Stability and approximator convergence in nonparametric nonlinear adaptive control," *IEEE Transactions on Neural Networks*, Vol. 9, pp. 1008–1020, 1998.

47. A. Feuer, and A. S. Morse, "Adaptive control of single-input, single-output linear systems," *IEEE Transactions on Automatic Control*, Vol. 23, pp. 557–569, 1978.

48. U. Forssell, and L. Ljung, "Closed-loop identification revisited," *Automatica*, Vol. 35, pp. 1215–1241, 1999.

49. A. L. Fradkov, and A. Yu. Pogromsky, *Introduction to Control of Oscillations and Chaos*, World Scientific, Singapore, 1998.

50. P. M. Frank, "Fault diagnosis in dynamic systems using analytical and knowledge-based redundancy: A survey and some new results," *Automatica*, Vol. 26, pp. 459–474, 1990.

51. R. Franke, "Scattered data interpolation: Tests of some methods," *Math. Comp.*, Vol. 38, pp. 181–199, 1982.

52. R. A. Freeman, and P. V. Kokotovic, *Robust Nonlinear Control Design*, Birkauser, Boston, 1996.

53. R. A. Freeman, M. Krstic, and P. V. Kokotovic, "Robustness of adaptive nonlinear control to bounded uncertainties," *Automatica*, Vol. 34, no. 10, pp. 1227–1230, 1998.

54. W. J. Freeman, "The physiology of perception," *Scientific American*, Vol. 264, no. 2, pp. 78–85, 1991.

55. K. S. Fu, "Learning control systems," in *Advances in Information Systems Science*, J. T. Tou, Ed., Plenum Press, New York, 1969.

56. K. S. Fu, "Learning control systems: Review and outlook," *IEEE Transactions on Automatic Control*, Vol. 15, pp. 210–221, April 1970.

57. K. S. Fu, "Learning control systems and intelligent control systems: An intersection of artificial intelligence and automatic control," *IEEE Transactions on Automatic Control*, Vol. 16, pp. 70–72, February 1971.

58. K. I. Funahashi, "On the approximate realization of continuous mappings by neural networks," *Neural Networks*, Vol. 2, pp. 183–192, 1989.

59. K. Funahashi, "On the approximate realization of continuous mappings by neural networks," *Neural Networks*, 1989.

60. F. Gandolfo, F. A. Mussa-Ivaldi, and E. Bizzi, "Motor learning by field approximation," *Proceedings of National Academy of Science*, Vol. 93, pp. 3843–3846, 1996.

61. J. P. Gauthier, H. Hammouri, and S. Othman, "A simple observer for nonlinear systems: Applications to bioreactors," *IEEE Transactions on Automatic Control*, Vol. 37, pp. 875–880, 1992.

62. J. P. Gauthier, and I. A. K. Kupka, *Deterministic Observation Theory and Applications*, Cambridge University Press, Cambridge, 2001.

63. S. S. Ge, T. H. Lee, and C. J. Harris, *Adaptive Neural Network Control of Robotic Manipulators*, World Scientific, London, 1998.

64. S. S. Ge, C. C. Hang, T. H. Lee, and T. Zhang, *Stable Adaptive Neural Network Control*, Kluwer Academic, Norwell, Massachusetts, 2001.

65. S. S. Ge, and C. Wang, "Direct adaptive NN control of a class of nonlinear systems," *IEEE Transactions on Neural Networks*, Vol. 13, no. 1, pp. 214–221, 2002.

66. S. S. Ge, and C. Wang, "Adaptive NN control of uncertain nonlinear pure-feedback systems," *Automatica*, Vol. 38, pp. 671–682, 2002.

67. S. S. Ge, and C. Wang, "Adaptive neural control of uncertain MIMO nonlinear systems," *IEEE Transactions on Neural Networks*, Vol 15, no. 3, pp. 674–692, 2004.

68. J. J. Gertler, "Survey of model-based failure detection and isolation in complex plants," *IEEE Control Systems Magazine*, Vol. 8, pp. 3–11, December 1988.

69. M. Gevers, and L. Ljung, "Optimal experiment designs with respect to the intended model application," *Automatica*, Vol. 22, pp. 543–554, 1986.

70. M. Gevers, "Identification for control: From the early achievements to the revival of experiment design," *European Journal of Control*, Vol. 11, pp. 1–18, 2005.

71. M. Gevers, "A personal view of the development of system identification," *IEEE Control Systems Magazine*, Vol. 12, pp. 93–105, December 2006.

72. S. Grossberg, "Some networks that can learn, remember, and reproduce any number of complicated space-time patterns," I, *Journal of Mathematics and Mechanics*, Vol. 19, pp. 53–91, 1969.

73. M. A. Cohen, and S. Grossberg, "Absolute stability and global pattern formation and parallel storage by competitive neural networks," *IEEE Transactions on Systems, Man, and Cybernectics*, Vol. 13, pp. 815–826, 1983.

74. S. Grossberg, "Nonlinear neural networks principles, mechanisms, and architectures," *Neural Networks*, Vol. 1, pp. 17–66, 1988.

75. M. Golubitsky, D. Luss, and S. H. Strogatz, eds., *Pattern Formation in Continuous and Coupled Systems: A Survey Volume*, Springer, New York, 1999.

76. J. Gong, and B. Yao, "Neural network adaptive robust control of nonlinear systems in semi-strict feedback form," *Automatica*, Vol. 37, pp. 1149–1160, 2001.

77. G. C. Goodwin, P. J. Ramadge, and P. E. Caines, "Discrete-time multi-variable adaptive control," *IEEE Transactions on Automatic Control*, Vol. 25, no. 3, pp. 449–456, 1980.

78. G. C. Goodwin, and K. C. Sin, *Adaptive Filtering Prediction and Control*, Prentice Hall, Englewood Cliffs, New Jersey, 1984.

79. G. C. Goodwin, and D. Q. Mayne, "A parameter estimation perspective of continuous time adaptive control," *Automatica*, Vol. 23, 1987.

80. D. Gorinevsky, "On the persistence of excitation in radial basis function network identification of nonlinear systems," *IEEE Transactions on Neural Networks*, Vol. 6, no. 5, pp. 1237–1244, 1995.

81. J. Guckenheimer, and P. Holmes, *Nonlinear Oscillations, Dynamical Systems, and Bifurcation- SOF Vector Fields*, Springer-Verlag, New York, 1983.

82. M. M. Gupta, and D. H. Rao, *Neuro-Control Systems: Theory and Applications*, IEEE Neural Networks Council, New York, 1994.

83. R. L. Hardy, "Multiquadric equations of topography and other irregular surfaces," *J. Geophys. Res.*, Vol. 76, pp. 1905–1915, 1971.

84. R. L. Hardy, "Theory and applications of the multiquadric-biharmonic method," *Comput. Math. Appl.*, Vol. 19, pp. 163–208, 1990.

85. S. Haykin, *Neural Networks: A Comprehensive Foundation*, 2nd ed., Prentice-Hall, Englewood Cliffs, New Jersey, 1999.

86. J. P. Hespanha, and A. S. Morse, "Scale-independent hysteresis switching," in *Hybrid Systems: Computation and Control*, F. W. Vaandrager, and J. H. van Schuppen, Eds., Lecture Notes in Computer Science, Vol. 1569, Springer, Berlin, pp. 117–122, 1999.

87. H. Hjalmarsson, M. Gevers, and F. De Bruyne, "For model-based control design, closed-loop identification gives better performance," *Automatica*, Vol. 32, pp. 1659–1673, 1996.

88. B. L. Ho, and R. E. Kalman, "Effective construction of linear state-variable models from input-output functions," *Regelungstechnik*, Vol. 12, pp. 545–548, 1965.

89. L. Holmstrom, and P. Koistinen, "Using additive noise in back-propagation training," *IEEE Transactions on Neural Networks*, Vol. 3, no. 1, pp. 24–38, 1992.

90. Y. Hong, J. Huang, and Y. Xu, "On an output feedback finite-time stabilization problem," *IEEE Transactions on Automatic Control*, Vol. 46, no. 2, pp. 305–309, 2001.

91. K. Hornik, M. Stinchcombe, and H. White, "Multilayer feedforward networks are universal approximators," *Neural Networks*, Vol. 2, pp. 359–366, 1989.

92. P. A. Ioannou, and J. Sun, *Robust Adaptive Control*, Prentice-Hall, Englewood Cliffs, New Jersey, 1995.

93. A. Isidori, *Nonlinear Control Systems*, Springer, Berlin, 1995.

94. A. Isidori, *Nonlinear Control Systems II*, Springer-Verlag, London, 1999.

95. A. K. Jain, R. P. W. Duin, and J. Mao, "Statistical pattern recognition: A review," *IEEE Transactions on Pattern Analysis and Machine Intelligence*, Vol. 22, no. 1, pp. 4–37, 2000.

96. M. Jankovic, "Adaptive nonlinear output feedback tracking with a partial high-gain observer and backstepping," *IEEE Transactions on Automatic Control*, Vol. 42, no. 1, pp. 106–113, 1997.

97. D. Jiang, and J. Wang, "On-line learning of dynamical systems in the presence of model mismatch and disturbances," *IEEE Transactions on Neural Networks*, Vol. 11, no. 6, pp. 1272–1283, 2000.

98. Z. P. Jiang, and I. M. Y. Mareels, "Small-gain control method for nonlinear cascaded systems with dynamic uncertainties," *IEEE Transactions on Automatic Control*, Vol. 42, no. 3, pp. 292–308, 1997.

99. Z. P. Jiang, and L. Praly, "Design of robust adaptive controllers for nonlinear systems with dynamic uncertainties," *Automatica*, Vol. 34, no. 7, pp. 825–840, 1998.

100. Z. P. Jiang, "A combined backstepping and small-gain approach to adaptive output feedback control," *Automatica*, Vol. 35, no. 6, pp. 1131–1139, June 1999.

101. M. I. Jordan, "Attractor dynamics and parallelism in a connectionist sequential machine," *Proceedings of the Eighth Annual Conference of the Cognitive Science Society*, Hillsdale, New Jersey, pp. 531–546, 1986.

102. A. Juditsky, H. Hjalmarsson, A. Benveniste, B. Delyon, L. Ljung, J. Sjoberg, and Q. Zhang, "Nonlinear black-box models in system identification: Mathematical foundations," *Automatica*, Vol. 31, pp. 1725–1750, 1995.

103. R. E. Kalman, and J. E. Bertram, "Control systems analysis and design via the 'second method' of Lyapunov," *Journal of Basic Engineering*, Vol. 82, pp. 371–392, 1960.

104. I. Kanellakopoulos, P. Kokotovic, and A. Morse, "Systematic design of adaptive controllers for feedback linearizable systems," *IEEE Transactions on Automatic Control*, Vol. 36, pp. 1241–1253, 1991.

105. T. Katayama, *Subspace Methods for System Identification*, Springer, New York, 2005.
106. O. Kaynak (Ed.), "Special issue on parameter adaptation and learning in computationally intelligent systems," *Int. J. Adaptive Control and Signal Processing*, Vol. 17, 2003.
107. N. Kazantzis, and C. Kravaris, "Nonlinear observer design using Lyapunov's auxiliary theorem," *Systems & Control Letters*, Vol. 34, pp. 241–247, 1998.
108. J. A. S. Kelso, *Dynamic Patterns: The Self-Organization of Brain and Behavior*, MIT Press, Cambridge, Massachusetts, 1995.
109. H. K. Khalil, and A. Saberi, "Adaptive stabilization of a class of nonlinear systems using high gain feedback," *IEEE Transactions on Automatic Control*, Vol. 32, no. 11, pp. 1031–1035, 1987.
110. H. K. Khalil, *Nonlinear Systems*, 2nd ed., Prentice Hall, Englewood Cliffs, New Jersey, 1996.
111. H. K. Khalil, *Nonlinear Systems*, 3rd ed., Prentice Hall, Englewood Cliffs, New Jersey, 2002.
112. Y. H. Kim, F. L. Lewis, and C. T. Abdallah, "A dynamic recurrent neural-network-based adaptive observer for a class of nonlinear systems," *Automatica*, Vol. 33, no. 8, pp. 1539–1543, 1997.
113. P. Kokotovic, and M. Arcak, "Constructive nonlinear control: A historical perspective," *Automatica*, Vol. 37, no. 5, pp. 637–662, May 2001.
114. E. B. Kosmatopoulos, M. M. Polycarpou, M. A. Christodoulou, and P. A. Ioannou, "High-order neural network structures for identification of dynamical systems," *IEEE Transactions on Neural Networks*, Vol. 6, no. 2, pp. 422–431, 1995.
115. E. B. Kosmatopoulos, M. A. Christodoulou, and P. A. Ioannou, "Dynamical neural networks that ensure exponential identification error convergence," *Neural Networks*, Vol. 10, no. 2, pp. 299–314, 1997.
116. S. R. Kou, D. L. Elliott, and T. J. Tarn, "Exponential observers for nonlinear dynamic systems," *Information and Control*, Vol. 29, pp. 204–216, 1975.
117. M. Krstic, I. Kanellakopoulos, and P. Kokotovic, "Adaptive nonlinear control without overparametrization," *Systems & Control Letters*, Vol. 19, pp. 177–185, 1992.
118. M. Krstic, P. V. Kokotovic, and I. Kanellakopoulos, "Transient performance improvement with a new class of adaptive controllers," *Systems & Control Letters*, Vol. 21, pp. 451–461, 1993.
119. M. Krstic, I. Kanellakopoulos, and P. Kokotovic, *Nonlinear and Adaptive Control Design*, John Wiley, New York, 1995.
120. Y. A. Kuznetsov, *Elements of Applied Bifurcation Theory*, 2nd ed, Springer, New York, 1998.
121. M. Krstic, D. Fontaine, P. Kokotovic, and J. Paduano, "Useful nonlinearities and global bifurcation control of jet engine stall and surge," *IEEE Transactions on Automatic Control*, Vol. 43, pp. 1739–1745, 1998.
122. M. Krstic, and H. Deng, *Stabilization of Nonlinear Uncertain Systems*, Springer-Verlag, New York, 1998.
123. A. J. Kurdila, F. J. Narcowich, and J. D. Ward, "Persistence of excitation in identification using radial basis function approximants," *SIAM Journal of Control and Optimization*, Vol. 33, no. 2, pp. 625–642, 1995.
124. C. Kwan, and F. L. Lewis, "Robust backstepping control of nonlinear systems using neural networks," *IEEE Transactions on Systems, Man and Cybernetics, Part A*, Vol. 30, no. 6, pp. 753–766, 2000.

125. I. D. Landau, *Adaptive Control: The Model Reference Approach*, Marcel Dekker, New York, 1979.

126. S. Lang, *Real Analysis*, Addison-Wesley, Reading, Massachusetts, 1983.

127. F. L. Lewis, A. Yesildirek, and K. Liu, "Multilayer neural-net robot controller with guaranteed tracking performance," *IEEE Transactions on Neural Networks*, Vol. 7, no. 2, pp. 388–398, 1996.

128. F. L. Lewis, S. Jagannathan, and A. Yeildirek, *Neural Network Control of Robot Manipulators and Nonlinear Systems*, Taylor & Francis, London, 1999.

129. F. L. Lewis, A. Yesildirek, and K. Liu, "Robust backstepping control of induction motors using neural networks," *IEEE Transactions on Neural Networks*, Vol. 11, no. 5, pp. 1178–1187, 2000.

130. L. L. Li, and D. H. Zhou, "Fast and robust fault diagnosis for a class of nonlinear systems: Detectability analysis," *Computers and Chemical Engineering*, pp. 2635–2646, July 2004.

131. D. Liberzon, *Switching in Systems and Control*, Birkhäuser, Boston, 2003.

132. J. S. Lin, and I. Kanellakopoulos, "Nonlinearities enhance parameter convergence in strict feedback systems," *IEEE Transactions on Automatic Control*, Vol. 44, pp. 89–94, 1999.

133. T. F. Liu, and C. Wang, "Learning from neural control of general Brunovsky systems," *Proceedings of the 21th IEEE International Symposium on Intelligent Control*, Munich, Germany, pp. 2366–2371, October 2006.

134. T. F. Liu, and C. Wang, "Learning from neural control of strict-feedback systems," *Proceedings of the 2007 IEEE International Conference on Control and Automation*, Guangzhou, China, pp. 636–641, June 2007.

135. L. Ljung, "On consistency and identifiability," *Mathematical Programming Study*, Vol. 5, pp. 169–190, 1976.

136. L. Ljung, "Convergence analysis of parametric identification methods," *IEEE Transactions on Automatic Control*, Vol. AC-23, pp. 770–783, October 1978.

137. L. Ljung and P. E. Caines, "Asymptotic normality of prediction error estimators for approximative system models," *Stochastics*, Vol. 3, pp. 29–46, 1979.

138. L. Ljung, "Asymptotic variance expressions for identified black-box transfer function models," *IEEE Transactions on Automatic Control*, Vol. AC-30, pp. 834–844, 1985.

139. L. Ljung, *System Identification: Theory for the User*, Prentice-Hall, Englewood Cliffs, New Jersey, 1987.

140. L. Ljung, *System Identification: Theory for the User*, 2nd ed., Prentice-Hall, Englewood Cliffs, New Jersey, 1999.

141. L. Ljung, "Challenges of non-linear system identification," Bode Lecture, *Proceedings of the 42nd IEEE Conference on Decision and Control*, Hawaii, December 2003.

142. E. N. Lorenz, "Deterministic non-periodic flow," *J. Atmos. Sci.* Vol. 20, pp. 130–141, 1963.

143. S. Lu, and T. Basar, "Robust nonlinear system identification using neural-network models," *IEEE Transactions on Neural Networks*, Vol. 9, no. 3, pp. 407–429, 1998.

144. W. R. Madych, and S. T. Nelson, "Multivariate interpolation and conditionally positive definite functions," *Approx. Theory Appl.*, Vol. 4, pp. 77–79, 1988.

145. W. R. Madych, and S. T. Nelson, "Multivariate interpolation and conditionally positive definite functions," II, *Math. Comp.*, Vol. 54, pp. 211–230, 1990.

146. R. Marino, and P. Tomei, "Global adaptive observers for nonlinear systems via filtered transformations," *IEEE Transactions on Automatic Control*, Vol. 37, pp. 1239–1245, August 1992.

147. R. Marino, and P. Tomei, *Nonlinear Adaptive Design: Geometric, Adaptive, and Robust*, Prentice Hall, London, 1995.

148. R. Marino, and P. Tomei, "Adaptive observers with arbitrary exponential rate of convergence for nonlinear systems," *IEEE Transactions on Automatic Control*, Vol. 40, pp. 1300–1304, July 1995.

149. R. Marino, and P. Tomei, "Robust adaptive state-feedback tracking for nonlinear systems," *IEEE Transactions on Automatic Control*, Vol. 43, pp. 84–89, 1998.

150. W. S. McCulloch, and W. Pitts, "A logical calculus of the ideas immanent in nervous activity," *Bull. Math. Biophys*, Vol. 5, pp. 115–133, 1943.

151. C. A. Micchelli, "Interpolation of scattered data: Distance matrices and conditionally positive definite functions," *Constructive Approximation*, Vol. 2, pp. 11–22, 1986.

152. R. H., Middleton, G. C. Goodwin, D. J. Hill, and D. Q. Mayne, "Design issues in adaptive control," *IEEE Transactions on Automatic Control*, Vol. 33, no. 1, pp. 50–58, 1988.

153. A. P. Morgan and K. S. Narendra, "On the stability of nonautonomous differential equations $\dot{x} = [A + B(t)]x$, with skew symmetric matrix $B(t)$," *SIAM Journal of Control and Optimization*, Vol. 15, no. 1, pp. 163–176, 1977.

154. J. Nakanishi, J. A. Farrell, and S. Schaal, "Composite adaptive control with locally weighted statistical learning," *Neural Networks*, Vol. 18, pp. 71–90, 2005.

155. F. J. Narcowich and J. D. Ward, "Norms of inverses and condition numbers for matrices associated with scattered data," *J. Approx. Theory*, Vol. 64, pp. 69–94, 1991.

156. F. J. Narcowich, and J. D. Ward, "Norms of inverses for matrices associated with scattered data," in *Curves and Surfaces*, P. J. Laurent, A. Le Mhaut, and L. L. Schumaker, Eds., Academic Press, Boston, 1991.

157. F. J. Narcowich, and J. D. Ward, "Norm estimates for the inverses of a general class of scattered-data radial-function interpolation matrices," *J. Approx. Theory*, Vol. 69, pp. 84–109, 1992.

158. F. J. Narcowich, R. Schaback, and J. D. Ward (1999), "Multilevel interpolation and approximation," *Appl. Comput. Harm. Analysis*, Vol. 7, pp. 243–261.

159. K. S. Narendra (Ed.), *Adaptive and Learning Systems: Theory and Applications*, Plenum Press, New York, 1986.

160. K. S. Narendra, and A. M. Annaswamy, "Persistent excitation of adaptive systems," *International Journal of Control*, Vol. 45, pp. 127–160, 1987.

161. K. S. Narendra and A. M. Annaswamy, *Stable Adaptive Systems*, Prentice-Hall, Englewood Cliffs, New Jersey, 1989.

162. K. S. Narendra, and K. Parthasarathy, "Identification and control of dynamic systems using neural networks," *IEEE Transactions on Neural Networks*, Vol. 1, no. 1, pp. 4–27, 1990.

163. K. S. Narendra, and S. Mukhopadhyay, "Intelligent control using neural networks," in *Intelligent Control Systems: Theory and Applications*, M. M. Gupta and N. K. Sinha, Eds., pp. 151–186, 1996.

164. K. S. Narendra, and J. Balakrishnan, "Adaptive control using multiple models," *IEEE Transactions on Automatic Control*, Vol. 42, no. 2, February 1997.

165. K. S. Narendra, and F. L. Lewis (Eds.), "Special issue on neural network feedback control," *Automatica*, Vol. 37, no. 8, 2001.

166. P. D. Neilson, M. D. Neilson, and N. J. O'Dwyer, "Adaptive optimal control of human tracking," *Motor Control and Sensory Motor Integration: Issues and Directions*, D. J. Glencross and J. P. Piek, Eds., Elsevier, Amsterdam, pp. 97–140, 1995.

167. O. Nelles, *Nonlinear System Identification: From Classical Approaches to Neural Networks and Fuzzy Models*, Springer, Berlin, 2001.

168. A. Newell and H. Simon, "Computer science and empirical inquiry," *Communications of the ACM*, pp. 113–126, 1975.

169. H. Nijmeijer, and T. I. Fossen (Eds.), *New Directions in Nonlinear Observer Design*, Springer-Verlag, London, 1999.

170. G. Nurnberger, *Approximation by Spline Functions*, Springer-Verlag, New York, 1989.

171. R. Ortega, "Some remarks on adaptive neuro-fuzzy systems," *Int. J. Adaptive Control and Signal Processing*, Vol. 10, pp. 79–83, 1996.

172. Z. Pan, and T. Basar, "Adaptive controller design for tracking and disturbance attenuation in parametric strict-feedback nonlinear systems," *IEEE Transactions on Automatic Control*, Vol. 43, no. 8, pp. 1066–1083, 1998.

173. E. Panteley, and A. Loría, "Uniform exponential stability for families of linear time-varying systems," *Proceedings of the 39th IEEE Conference on Decision and Control*, Sydney, Australia, December, 2000.

174. J. Park, and I. W. Sandberg, "Universal approximation using radial-basis-function networks," *Neural Computation*, Vol. 3, pp. 246–257, 1991.

175. J. H. Park, S. H. Huh, S. H. Kim, S. J. Seo, and G. T. Park, "Direct adaptive controller for nonaffine nonlinear systems using self-structuring neural networks," *IEEE Transactions on Neural Networks*, Vol. 16, no. 2, pp. 414–422, 2005.

176. R. J. Patton, R. N. Clark, and P. M. Frank, *Issues of Fault Diagnosis for Dynamic Systems*, Springer, Berlin, 2000.

177. T. Poggio, and F. Girosi, *A Theory of Networks for Approximating and Learning*, A. I. Memo No. 1140, Artificial Intelligence Laboratory, Massachusetts Institute of Technology, Cambridge, 1989.

178. T. Poggio, and F. Girosi, "Networks for approximation and learning," *Proceedings of the IEEE*, Vol. 79, pp. 1481–1497, 1990.

179. M. M. Polycarpou, and P. A. Ioannou, "Modeling, identification and stable adaptive control of continuous-time nonlinear dynamical systems using neural networks," *Proceedings of American Control Conference*, Boston, pp. 36–40, 1992.

180. M. M. Polycarpou, "Stable adaptive neural control scheme for nonlinear systems," *IEEE Transactions on Automatic Control*, Vol. 41, no. 3, pp. 447–451, 1996.

181. M. M. Polycarpou, and M. J. Mears, "Stable adaptive tracking of uncertain systems using nonlinearly parametrized on-line approximators," *International Journal of Control*, Vol. 70, no. 3, pp. 363–384, 1998.

182. M. J. D. Powell, "Radial basis functions for multivariable approximation," in *Algorithms for Approximation*, J. C. Mason and M. G. Cox, Eds., Oxford University Press, Oxford, 1987.

183. M. J. D. Powell, "The theory of radial basis function approximation in 1990," in *Advances in Numerical Analysis II: Wavelets, Subdivision, Algorithms, and Radial Basis Functions*, W. A. Light, Ed., Oxford University Press, Oxford, pp. 105–210, 1992.

184. J. Protz, and J. Paduano, "Rotating stall and surge: Alternate modeling and control concepts," *Proceedings of the IEEE International Conference on Control Applications*, Hartford, pp. 866–873, 1997.

185. Z. Qu, *Robust Control of Nonlinear Uncertain Systems*, John Wiley & Sons, New York, 1998.

186. O. E. Rössler, "An equation for continuous chaos," *Physica Letters*, Vol. 57A, pp. 397–398, 1976.

187. M. I. Rabinovich, A. B. Ezersky, and P. D. Weidman, *The Dynamics of Patterns*, World Scientific, Singapore, 2000.

188. R. Rajamani, "Observer for Lipschitz nonlinear systems," *IEEE Transactions on Automatic Control*, Vol. 43, pp. 397–401, March 1998.

189. J. R. Rice, *The Approximation of Functions*, Addison-Wesley, Reading, Massachusetts, 1964.

190. G. A. Rovithakis, and M. A. Christodoulou, "Adaptive control of unknown plants using dynamical neural networks," *IEEE Trans. Syst., Man, Cybern.*, Vol. 24, pp. 400–412, March 1994.

191. G. A. Rovithakis, "Tracking control of multi-input affine nonlinear dynamical systems with unknown nonlinearities using dynamical neural networks," *IEEE Trans. Syst., Man, Cybern.*, Vol. 29, no. 2, pp. 179–189, 1999.

192. J. A. Ruiz Vargas, and E. M. Hemerly, "Adaptive observers for unknown general nonlinear systems," *IEEE Transactions on Systems, Man, and Cybernetics, Part B: Cybernetics*, Vol. 31, no. 5, pp. 683–690, October 2001.

193. A. Sanfeliu, et al., "Graph-based representations and techniques for image processing and image analysis," *Pattern Recognition*, Vol. 35, no. 3, pp. 639–650, 2002.

194. R. M. Sanner, and J. E. Slotine, "Stable recursive identification using radial basis function networks," in *Proceedings of American Control Conference*, Vol. 3, Chicago, pp. 1829–1833, 1992.

195. R. M. Sanner, and J. E. Slotine, "Gaussian networks for direct adaptive control," *IEEE Transactions on Neural Networks*, Vol. 3, no. 6, pp. 837–863, 1992.

196. R. M. Sanner, and M. Kosha, "A mathematical model of the adaptive control of human arm motions," *Biological Cybernetics*, Vol. 80, pp. 369–382, 1999.

197. G. N. Saridis, "Toward the realization of intelligent controls", *Proc. IEE*, Vol. 67, no. 8, August 1979.

198. G. Saridis, "Intelligent robotic control," *IEEE Transactions on Automatic Control*, Vol. 28, pp. 547–557, May 1983.

199. S. S. Sastry, and M. Bodson, *Adaptive Control: Stability, Convergence, and Robustness*, Prentice-Hall, Englewood Cliffs, New Jersey, 1989.

200. S. S. Sastry, and A. Isidori, "Adaptive control of linearizable systems," *IEEE Transactions on Automatic Control*, Vol. 34, no. 11, pp. 1123–1131, 1989.

201. I. J. Schoenberg, "Metric spaces and completely monotone functions," *Annals of Mathematics*, Vol. 39, pp. 811–841, 1938.

202. O. Seungrohk, and H. K. Khalil, "Nonlinear output-feedback tracking using high-gain observer and variable structure control," *Automatica*, Vol. 33, no. 10, pp. 1845–1856, 1997.

203. R. Sepulchre, M. Jankovic, and P. V. Kokotovic, *Constructive Nonlinear Control*, Springer-Verlag, London, 1997.

204. R. Shadmehr, and F. A. Mussa-Ivaldi, "Adaptive representation of dynamics during learning of a motor task," *Journal of Neuroscience*, Vol. 14, no. 5, pp. 3208–3224, May 1994.

205. C. H. Shea, W. L. Shebilske, and S. Worchel, *Motor Learning and Control*, Prentice Hall, Englewood Cliffs, New Jersey, 1993.

206. L. P. Shilnikov, et al., *Methods of Qualitative Theory in Nonlinear Dynamics, Part I*, World Scientific, Singapore, 2001.

207. L. P. Shilnikov, et al., *Methods of Qualitative Theory in Nonlinear Dynamics, Part II*, World Scientific, Singapore, 2001.

208. A. K. Sinha, "Power system security assessment using pattern recognition and fuzzy estimation," *International Journal of Electrical Power & Energy Systems*, Vol. 17, no. 1, pp. 11–19, 1995.

209. J. Sjoberg, Q. Zhang, L. Ljung, A. Benveniste, B. Deylon, P.-Y. Glorennec, H. Hjalmarsson, and A. Juditsky, "Nonlinear black-box modeling in system identification: A unified overview," *Automatica*, Vol. 31, no. 12, pp. 1691–1724, December 1995.

210. C. A. Skarda, and W. J. Freeman, "How brains make chaos in order to make sense of the world," *Behavioral and Brain Sciences*, Vol. 10, pp. 161–195, 1987.

211. J. J. Slotine, and W. Li, *Applied Nonlinear Control*, Prentice Hall, Englewood Cliffs, New Jersey, 1991.

212. T. Soderstrom, and P. Stoica, *System Identification*, Prentice-Hall, Hemel Hempstead, Hertfordshire, UK, 1989.

213. E. D. Sontag, and Y. Wang, "On characterizations of the input-to-state stability property," *Systems & Control Letters*, Vol. 24, no. 5, pp. 351–359, 1995.

214. E. Sontag, Some topics in neural networks and control, Rutgers University, Report No. LS93-02, July 1993.

215. E. D. Sontag, and Y. Wang, "New characterizations of input-to-state stability," *IEEE Transactions on Automatic Control*, Vol. 41, no. 9, pp. 1283–1294, 1996.

216. J. T. Spooner, and K. M. Passino, "Stable adaptive control using fuzzy systems and neural networks," *IEEE Transactions on Fuzzy Systems*, Vol. 4, no. 3, pp. 339–359, 1996.

217. D. W. Tank, and J. J. Hopfield, "Neural computation by concentrating information in time," *Proceedings of the National Academy of Science*, Vol. 84, pp. 1896–1900, 1987.

218. D. Taylor, P. V. Kokotovic, R. Marino, and I. Kanellakopoulos, "Adaptive regulation of nonlinear systems with unmodeled dynamics," *IEEE Transactions on Automatic Control*, Vol. 34, pp. 405–412, 1989.

219. A. R. Teel, "Nonlinear small gain theorem for the analysis of control systems with saturation," *IEEE Transactions on Automatic Control*, Vol. 41, no. 9, pp. 1256–1270, 1996.

220. F. E. Thau, "Observing the state of non-linear dynamic systems," *Int. J. of Control*, Vol. 17, no. 3, pp. 471–479, 1973.

221. N. F. Thornhill, B. Huang, and H. Zhang, "Detection of multiple oscillations in control loops," *Journal of Process Control*, Vol. 13, no. 1, pp. 91–100, February 2003.

222. K. A. Thoroughman, and R. Shadmehr, "Learning of action through adaptive combination of motor primitives," *Nature*, Vol. 407, pp. 742–747, 2002.

223. A. B. Trunov, and M. M. Polycarpou, "Automated fault diagnosis in nonlinear multivariable systems using a learning methodology," *IEEE Transactions on Neural Networks*, Vol. 11, pp. 91–101, Feberuary 2000.

224. J. Tsinias, "Observer design for nonlinear systems," *Systems & Control Letters*, Vol. 13, p. 135, 1989.

225. J. Tsinias, "Further results on the observer design problem," *Systems & Control Letters*, Vol. 14, p. 411, 1990.

226. Tsypkin, Y. Z., *Adaptation and Learning in Automatic Systems*, Academic Press, New York, 1971.

227. B. van der Pol, "Forced oscillations in a circuit with nonlinear resistance (reception with reactive triode)," *Philosophical Magazine*, Vol. 7, pp. 65–80, 1927.

228. T. van Gelder, and R. Port, "It's about time: An overview of the dynamical approach to cognition," in *Mind as Motion: Explorations in the Dynamics of Cognition*, MIT Press, Cambridge, Massachusetts, 1995.

229. V. N. Vapnik, *Statistical Learning Theory*, John Wiley, New York, 1998.

230. V. N. Vapnik, *The Nature of Statistical Learning Theory*, 2nd ed., Springer, New York, 2000.

231. V. Venkatasubramanian, R. Rengaswamy, K. Yin, and S. N. Kavuri, "A review of process fault detection and diagnosis: Part I: Quantitative model-based methods," *Computers & Chemical Engineering*, Vol. 27, no. 3, pp. 293–311, March 2003.

232. V. Venkatasubramanian, R. Rengaswamy, and S. N. Kavuri, "A review of process fault detection and diagnosis: Part II: Qualitative models and search strategies," *Computers & Chemical Engineering*, Vol. 27, no. 3, pp. 313–326, March 2003.

233. M. Vidyasagar, *A Theory of Learning and Generalization: With Applications to Neural Networks and Control Systems*, Springer, London, 1997.

234. M. Vidyasagar, "Randomized algorithms for robust controller synthesis using statistical learning theory," *Automatica*, Vol. 37, pp. 1515–1528, 2001.

235. B. Wahlberg, and L. Ljung, "Design variables for bias distribution in transfer function estimation," *IEEE Transactions on Automatic Control*, Vol. AC-31, pp. 134–144, 1986.

236. A. Waibel, T. Hanazawa, G. Hinton, and K. Shikano, "Phoneme recognition using time-delay neural networks," *IEEE Transactions on Acoustics, Speech and Signal Processing*, Vol. 37, no. 3, pp. 328–339, 1989.

237. C. Wang, G. Chen, and S. S. Ge, "Smart neural control of uncertain nonlinear systems," *International Journal on Adaptive Control and Signal Processing—special issue on parameter adaptation and learning in computationally intelligent systems* (invited paper), Vol. 17, pp. 467–488, 2003.

238. C. Wang, D. J. Hill, and G. Chen, "Deterministic learning of nonlinear dynamical systems," *Proceedings of the 18th IEEE International Symposium on Intelligent Control*, pp. 87–92, Houston, Texas, October 2003.

239. C. Wang, G. Chen, and D. J. Hill, "Dynamical pattern classification," *IEEE Conference on Intelligent Automation*, Hong Kong, December 2003.

240. C. Wang, and D. J. Hill, "Learning from direct adaptive neural control," *5th Asian Control Conference*, Vol. 1, pp. 674–681, Melbourne, Australia, July 2004.

241. C. Wang, and D. J. Hill, "Deterministic learning from state observation," *Proceedings of the 23rd Chinese Control Conference*, Wuxi, China, August 10–13, 2004.

242. C. Wang, and D. J. Hill, "Persistence of excitation, RBF approximation and periodic orbits," *International Conference on Control and Automation*, Vol. 1, pp. 547–552, Budapest, Hungary, June 2005.

243. C. Wang, and D. J. Hill, "Learning from neural control," *IEEE Transactions on Neural Networks*, Vol. 17, no. 1, pp. 130–146, 2006.

244. C. Wang, and D. J. Hill, "Deterministic learning and rapid dynamical pattern recognition," *IEEE Transactions on Neural Networks*, Vol. 18, pp. 617–630, 2007.

245. C. Wang, D. J. Hill, S. S. Ge, and G. Chen, "An ISS-modular approach for adaptive neural control of pure-feedback systems," *Automatica*, Vol. 42, pp. 723–731, 2006.

246. C. Wang, C.-H. Wang, and S. Song, "An RBFN-based observer for nonlinear systems via deterministic learning," *2006 IEEE International Symposium on Intelligent Control*, pp. 2360–2365, Munich, Germany, October 2006.

247. C. Wang, T. Liu, and C.-H. Wang, "Deterministic learning and pattern-based NN control," *2007 IEEE Multi-conference on Systems and Control*, Singapore, October 2007.

248. C. Wang, C.-H. Wang, and S. Song, "Rapid recognition of dynamical patterns via deterministic learning and state observation," *2007 IEEE Multi-Conference on Systems and Control*, Singapore, October 2007.

249. D. L. Wang et al. (Eds.), "Special issue on temporal coding for neural information processing," *IEEE Transactions on Neural Networks*, Vol. 15, no. 5, September 2004.

250. D. Wang, and J. Huang, "Neural network-based adaptive dynamic surface control for a class of uncertain nonlinear systems in strict-feedback form," *IEEE Transactions on Neural Networks*, Vol. 16, no. 1, pp. 195–202, 2005.

251. D. L. Wang, and M. A. Arbib, "Complex temporal sequence learning based on short-term memory," *Proceedings of the IEEE*, Vol 78, pp. 1536–1543, 1990.

252. L. X. Wang, *Adaptive Fuzzy Systems and Control: Design and Analysis*, Prentice-Hall, Englewood Cliffs, New Jersey, 1994.

253. S. Weaver, L. Baird, and M. Polycarpou, "An analytical framework for local feedforward networks," *IEEE Transactions on Neural Networks*, Vol. 9, pp. 473–482, May 1998.

254. A. R. Webb, *Statistical Pattern Recognition*, 2nd ed., John Wiley & Sons, New York, 2002.

255. E. T. Whittaker, and G. N. Watson, *A Course in Modern Analysis*, 4th ed., Cambridge University Press, Cambridge, 1990.

256. D. V. Widder, *The Laplace Transform*, Princeton University Press, 1946.

257. P. H. Winston, *Artificial Intelligence*, 3rd ed., Addison-Wesley, Reading, Massachusetts, 1992.

258. C. Xia, J. Howell, and N. F. Thornhill, "Detecting and isolating multiple plant-wide oscillations via spectral independent component analysis," *Automatica*, Vol. 41, no. 12, pp. 2067–2075, December 2005.

259. X.-H. Xia and W.-B. Gao, "Non-linear observer design by observer canonical forms," *Int. J. Control*, Vol. 47, pp. 1081–1100, 1988.

260. B. Yao, and M. Tomizuka, "Adaptive robust control of SISO nonlinear systems in a semi-strict feedback form," *Automatica*, Vol. 33, no. 5, pp. 893–900, 1997.

261. B. Yegnanarayana, *Artificial Neural Networks*, Prentice-Hall of India, New Delhi, 1999.

262. A. Yesidirek, and F. L. Lewis, "Feedback linearization using neural networks," *Automatica*, Vol. 31, no. 11, pp. 1659–1664, 1995.

263. J. S.-C. Yuan, and W. M. Wonham, "Probing signals for model reference identification," *IEEE Transactions on Automatic Control*, Vol. 22, pp. 530–538, 1977.

264. M. Zeitz, "The extended Luenberger observer for nonlinear systems," *Systems & Control Letters*, Vol. 9, pp. 149–156, 1987.

265. T. Zhang, S. S. Ge, and C. C. Hang, "Design and performance analysis of a direct adaptive controller for nonlinear systems," *Automatica*, Vol. 35, pp. 1809–1817, 1999.

266. T. Zhang, S. S. Ge, and C. C. Hang, "Adaptive neural network control for strict-feedback nonlinear systems using backstepping design," *Automatica*, Vol. 36, pp. 1835–1846, 2000.

267. X. D. Zhang, M. M. Polycarpou, and T. Parisini, "A robust detection and isolation scheme for abrupt and incipient faults in nonlinear systems," *IEEE Transactions on Automatic Control*, Vol. 47, no. 4, pp. 576–593, April 2002.

268. Y. Zhang, P. A. Ioannou, and C.-C. Chien, "Parameter convergence of a new class of adaptive controllers," *IEEE Transactions on Automatic Control*, Vol. 41, pp. 1489–1493, 1996.
269. Y. Zhang, P. Y. Peng, and Z. P. Jiang, "Stable neural controller design for unknown nonlinear systems using backstepping," *IEEE Transactions on Neural Networks*, Vol. 11, no. 6, pp. 1347–1359, 2000.

Index

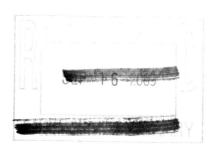